FOUR DAN SCULPTORS

CONTINUITY AND CHANGE

FOUR DAN SCULPTORS
CONTINUITY AND CHANGE

BARBARA C. JOHNSON

THE FINE ARTS MUSEUMS OF SAN FRANCISCO

DISTRIBUTED BY
THE UNIVERSITY OF CHICAGO PRESS

FOUR DAN SCULPTORS
CONTINUITY AND CHANGE

The Fine Arts Museums of San Francisco
M. H. de Young Memorial Museum
20 September 1986–1 February 1987

Honolulu Academy of Arts
Honolulu, Hawaii

This exhibition is supported in part by a grant from the National Endowment for the Arts and a contribution from the Mrs. Paul L. Wattis Exhibition Fund.

Published by The Fine Arts Museums of San Francisco, 1986.

Translations are by the author unless otherwise noted.

Photo Credits: Reproductions of works in the exhibition are by permission of the lenders. Photographs of the following cat. nos. are by: Michael Venera, 1; Joseph McDonald, 2, 4, 6, 12, 13, 21, 22, 26, 27, 29-31, 33, 37, 41-45; Hillel Burger, 3, 5, 32, 34, 35; Judy Cooper, 7; F. P. Orchard, 8, 36; Isabelle Wettstein and Brigette Kamerer-Kauf, 9, 14; Musée National des Arts Africains et Océaniens, 10; Igor Delmas, 11 and frontispiece; Rick Stafford, 15; Roy Trahan, 16; Fred Schock, 17-20; P. Horner, 23; Eugene R. Prince, 38; Margret Landsberg, 39; and Vic Krantz, 40. Photographs of the figure illustrations are by: Thomas K. Seligman, 1, 18; William C. Siegmann, cover, 2, 3, 5, 23, 31, 33, 34, 36-38; Barbara C. Johnson, 4, 9, 24, 29, 30, 35, 39-41, 44-46; Hillel Burger, 6; Peter Kerser, 7, 32, 42; Charles Uht, 8; Eugene R. Prince, 10; F. P. Orchard, 11; Charles Miller, 12; Hans Himmelheber, 13, 21, 22; Joseph McDonald, 15, 17; George Schwab, 19, 20; and Eberhard Fischer, 27, 28.

Front cover: Dancing *Kagle*, named Slü, seen in Nyor Diaple, February 1986. Background: detail, cat. no. 1.

Back cover: cat. no. 1.

Frontispiece: Figure of mother with baby on back. Jacques Kerchache collection, Paris, cat. no. 11.

This catalogue was produced by the Publications Department of The Fine Arts Museums of San Francisco. Ann Heath Karlstrom, Publications Manager. Karen Kevorkian, Editor. Designed by Ed Marquand, Ed Marquand Book Design, Seattle. Photocomposed by The Type Gallery, Seattle.

Printed in Hong Kong.

Johnson, Barbara C.
Four Dan Sculptors.
lll5Catalogue of an exhibition held at The Fine Arts Museums of San Francisco, M.H. de Young Memorial Museum, 20 Sept. 1986–1 Feb. 1987.

Bibliography: p. 101
1. Gere (African people)—Masks—Exhibitions. 2. Woodcarving, Primitive—Liberia—Exhibitions. 3. Brasswork—Liberia—Exhibitions. 4. Gere (African people)—Social life and customs. I. Fine Arts Museums of San Francisco.
NK9789.6.L5J64 1986   730'.089963   86-14308
ISBN 0-88401-049-X

Contents

Foreword

The Fine Arts Museums of San Francisco are pleased to present in this exhibition a close look at four individual artists of the Dan people of Liberia. The work of these three wood-carvers and one figurative brass-caster spans three generations. Considering the work of each in its cultural context permits a glimpse of a West African artist's relationship to his patrons, and an in-depth look at the Dan art tradition, which has long been admired in collections of African art.

This catalogue was written by Barbara C. Johnson, guest curator of the exhibition. It was begun as a project in fulfillment of a master of arts degree from San Francisco State University in 1984, and has grown as knowledge was gained from two research visits to the Dan people of Liberia, and as suggestions and advice were generously offered. Particularly helpful were Warren d'Azevedo, Professor of Anthropology, University of Nevada at Reno; Eberhard Fischer, Director, Rietberg Museum in Zürich; Svend Holsoe, Associate Professor of Anthropology, University of Delaware; Thomas K. Seligman, Deputy Director for Education and Exhibitions and Curator in Charge of the Department of Africa, Oceania, and the Americas, The Fine Arts Museums of San Francisco; William C. Siegmann, Fulbright Fellow, United States Education and Cultural Foundation, Monrovia; and Sylvia Williams, Director, National Museum of African Art, Smithsonian Institution.

The exhibition borrows from a number of collections in the United States and Europe. I want to thank the Lowie Museum of Anthropology of the University of California, Berkeley; the New Orleans Museum of Art; the National Museum of Natural History, Smithsonian Institution; the Musée National des Arts Africains et Océaniens in Paris; and the Peabody Museum of Archaeology and Ethnology, Harvard University; as well as the many private collectors who have generously allowed us to include parts of their collections in this exhibition and catalogue. I am particularly grateful to Mrs. Paul L. Wattis and to the National Endowment for the Arts for support of this exhibition and catalogue.

Ian McKibbin White
Director of Museums

Acknowledgments

I am grateful to many people who have helped me with this project. First, thanks are due William C. Siegmann, who helped me to begin this project, consistently giving me the direction I needed and accompanying me on two field research trips to the Dan villages in Nimba County. I am also grateful to San Francisco State University, which awarded me a master of arts degree in ethnic arts museology after I completed an earlier version of this catalogue, and to the faculty members who helped me with it, particularly David P. Gamble, Robin F. Wells, Marian Bernstein, and Judith Bettelheim.

In addition I owe a special debt to members of the staff of The Fine Arts Museums of San Francisco for their effort, necessary for the realization of this project. Thomas K. Seligman, Deputy Director for Education and Exhibitions and Curator in Charge of the Department of Africa, Oceania, and the Americas, consistently offered encouragement and guidance, as did Kathleen Berrin, Curator. This manuscript benefits from careful work by Ann Heath Karlstrom, Publications Manager, and Karen Kevorkian, Editor. Thanks also are due Debra Pughe, Exhibitions Manager; Paula March and Lisa Karplus, Registration; Antonette De Vito, Development Associate; Kathy Baldwin, Volunteer Coordinator; Pamela Forbes, Editor and Production Manager of *Triptych*; Renée Beller Dreyfus, Associate Curator for Education and Interpretation; Max Chance, Exhibition Architect; Ron Rick, Senior Graphic Designer; Michael Sandgren, Packer; Couric Payne, Bookshops Manager; William White, Technical Coordinator; and the crew of technicians who installed the exhibition.

A debt of thanks is due as well to others who provided help on the catalogue, among them Margarita Herbert, Pauline Jacobson, and Eva Lischer for translations important to full understanding of the literature; Ed Marquand Communication Design for the design and production of the catalogue; and the consultants who advised and helped me with the information in this manuscript. They were, in particular, Warren d'Azevedo, Eberhard Fischer, Svend Holsoe, Thomas K. Seligman, William C. Siegmann, and Sylvia Williams. It is also true that this work would have been impossible without the writings of previous students of the Dan and their neighbors.

Without the many individuals and institutions who were willing to lend us the objects for this exhibition, we would not have one. I am grateful to all of them. Above all thanks are due to the Mrs. Paul L. Wattis Exhibition Fund and to the National Endowment for the Arts, whose generous support made the exhibition possible.

None of this would have been possible without the Dan people who received me with gracious hospitality in the towns of Nuopie, Kpeaple, Belewale, Yuopie Old Town, Butuo, Tapita, Toway Town, Gaple, Gbanwea, Nyor Diaple, Barluople, Beple, and Kanple. I am grateful particularly to the carvers who shared much information and made this work possible. Thanks are due George Wowoa Tabmen for sharing information on Dan culture, and very special thanks are due Peter Kerser, who functioned as Dan translator and an excellent research assistant. I am also grateful to the government of Liberia for permission to pursue research in the country in 1983 and 1986.

And finally a word of deeply felt gratitude is reserved for my family: my husband, Robert, who has always encouraged me and who has been willing to manage the household alone while I have pursued research on another continent, and my three children, who are only now beginning to understand a little of their mother's enthusiasm.

Barbara C. Johnson
Guest Curator

Preface

As Tom Seligman ably points out in his introduction to this work, much African art is presented to us anonymously. Hoping to redress this imbalance, I first studied the published writings of previous students of the Dan and their neighbors. Without the writings of Donner, Harley, Schwab, Vandenhoute, Holas, Siegmann, Tabmen, and especially Himmelheber and Fischer, among others, this work would have been impossible. These writers inevitably led to the source of the investigation, the Dan people themselves—in particular the carvers—and to people who remembered the carvers and brass-casters who had died. I made two short research trips to Dan country in northeast Liberia in 1983 and 1986. Information obtained on these visits added to that published earlier has led to this exhibition and publication, the principal aim of which is to introduce four individual Dan sculptors and their works.

In the Dan culture it is considered important to become the best at something. Every young Dan man and woman seeks to make a name, to achieve a reputation, *tin,* by excelling in a chosen endeavor. *Tin kadu* is the name for the person who seeks this kind of prestige (Fischer 1978, 16). A man may strive to be best at clearing the fields for planting, at wrestling, hunting, drumming, dancing, or wood carving, to name a few possibilities. A woman's choices include basket or pottery making, midwifery, and being the most hospitable woman of her village quarter. Possibly because of this pervasive competitive bent in the Dan culture, much of its sculpture is of the highest quality. The wood-carvers I interviewed had worked hard to become the best in their areas, and typically spoke in a self-praising manner to further establish that reputation. This is the expected behavior of a successful Dan person.

The Dan[1] are part of a larger cluster of different ethnic groups who speak languages from two major linguistic families, Kru and Mande. Included are the We,[2] Mano, Kono, Mau, Toura, Diomande, Wobe, Niabwa, Bete, Bassa, and Grebo. These groups, which inhabit parts of the Ivory Coast, Liberia, and Guinea, share certain features of social, political, and religious organization and belief. Among these features are similar sculptural forms and patterns of use. Despite variations in regional substyles, these objects represent a fundamental cultural unity that may be perceived throughout the region.

Regional substyles often overlap ethnolinguistic boundaries. This is particularly the case along the border between the southern Dan and western We, a region that has seen a great deal of cultural interchange. Many of the people in this area are bilingual, and intermarriage is quite common. It is generally conceded that the Dan are the prominent cultural force in this particular subregion. The evidence for this view notes the same sculptural forms in this border area that also are found throughout the Dan region; among the We the forms are found mainly in their westernmost region, adjacent to the Dan.

For this reason I chose to focus on the Dan although two of the carvers presented here were, in fact, born of We-speaking fathers. According to African tradition, they

---

[1] Also called Gio and Yacuba in conventional local parlance.

[2] Pronounced as the English "way." Also called Kran, Guéré, Gere, and Ngere. We is the name generally agreed upon by orthographers.

should thus identify with a We ethnicity. Their association with the Dan, however, is a very strong one. Both We carvers of this exhibition were bilingual and spoke fluent Dan. In addition, both worked primarily for commissions in the Dan villages of the region, and one worked extensively for many Dan chiefs. Even when they carved for their own We people, the forms they carved were apparently indistinguishable from those of the Dan.

Introduction
Four Dan Sculptors

Since the earliest western contacts with Africa in the fifteenth century, finely crafted objects have been brought out by explorers, traders, and missionaries. These objects generally were viewed as curios by westerners—examples of the exotic ways of strange people. The ethnic group from which the object came was poorly known, if at all. Only in the late nineteenth and early twentieth centuries did westerners develop interest in the people and cultures of the accessible parts of Africa that had been explored and were under colonial control.

Slowly, anthropologists began to study "tribes," researching their ways of life. Objects that had been admired only for their aesthetic qualities and fine craftsmanship began to be seen as functional items in the repertory of a culture. Their use, and how that use fit into the cultural framework, was what was important. Even after African art was discovered by avant-garde European artists in the early part of this century, the African objects so discovered were seen by these western artists only as sources of inspiration and confirmation of new aesthetic directions. It is curious that while there was recognition and appreciation on this aesthetic level, there seems to have been no interest by the western artist in learning who made the objects they so admired. Knowing the geographical region or "tribe" seems to have been more than sufficient, and the customs of the people who made the art were of little interest.

As the discipline of anthropology matured, and as several African societies became better known, it became clear that the cultural role of the artist, and the artist as an individual, were important to understand in their own right and in relation to social activities. As did anthropologists, art historians became interested in African art and conducted field investigations focusing on the art and the artist. However, in the last two decades a substantial body of work emerged that focused on the individual artist. Scholars such as Fagg, Bascom, Thompson, Fernandez, Himmelheber, Fischer, and Vogel have researched living artists and told their stories, while also broadening our understanding of the general role of the artist in his or her society. Nevertheless, as late as 1973, in the preface to a major book on the African artist, Alan Merriam and Roy Sieber lament, "While the literature abounds with technical as well as historical descriptions of aspects of music, art, oral literature and even to some extent dance and drama, we know virtually nothing of the artist as a creative member of society…" (d'Azevedo 1973, xx).

Among the earliest research on African artists is that conducted by Hans Himmelheber and his son Eberhard Fischer among the Dan of Liberia and the Ivory Coast. Many have benefited from their work, which helped provide a focus on Liberian art and allowed broader recognition and understanding of its significant aesthetic qualities. *Four Dan Sculptors* is certainly within the mainstream of this inquiry, and would not have been possible without their efforts.

Certain common themes emerge from all the research that has been done, specifically from that on Liberia drawing on the work by Himmelheber, Fischer, d'Azevedo,

Fig. 1   Varney Massalay and Momo Fina Massalay
Kendeja, Liberia, 1981

Siegmann, and Johnson. It is not surprising to find that the artist is a product of his society, and that the society is as ambivalent about the artist as the artist is ambivalent about himself. Daniel Biebuyck says, "We may conclude that there is a close, reciprocal bond between the artist and his community, which both compels him to do certain things and allows him to do other things in his own way" (1969, 10). Focusing on the process of creation, James Fernandez indicates that "the carver has the prerogatives of his creation, but he must still reach some accommodation with his critics who not infrequently consider themselves the final cause of the work" (d'Azevedo 1973, 203).

Among the themes that have emerged, it is clear that the quest for status and having a good name is highly prized by artists of different ethnic groups in Liberia. As exemplified by the Gola in the western part of Liberia, "To be renowned is to have one's ways and one's handiwork recognized far and wide, to have one's name attached to one's acts and things, to become legend" (d'Azevedo 1973, 311). Similarly, the Dan in the East share this desire, as Himmelheber and Fischer indicate: "The Dan man or woman strives to obtain distinction in the village by fulfilling a role appropriate to his character and abilities, trying to make a good name for himself" (Fischer and Himmelheber 1984, 4). I found while working in 1980 with two Vai artists, Momo Fina Massalay and his son Varney Massalay, that Momo also agreed with this attitude. "No, I carve for my name. I'm not carving for the name of money," was his observation, even though his entire recent carving career had been directed to a western audience of missionaries, miners, and Peace Corps volunteers (fig. 1).

Another common feature in the research is the artists' frequent inspiration from dreams. In these they receive, or learn of, a spirit helper (*Neme* among the Vai and Gola, *dü* among the Dan). It is these spirit helpers themselves who inspire the artist, but it is also the existence of these metaphysical beings that causes the artist to be viewed by his society with ambivalence, suspicion, and fear because of the power obtained from the other world. However, the artist also is viewed with respect and admiration because he produces the spiritually powerful, beautiful, and useful objects the society requires. The artist thus is caught in a conflict because of his need to create, to be different, and to be a respected individual, but also because of his need to gain the desired and necessary respect from members of his society. The artist's classic dilemma is whether to serve art or society, each reality imposing conflicting demands.

It is commonly acknowledged that one becomes an artist by being given the talent by God. As Zon, a Dan carver interviewed by both Fischer and Johnson, expressed it, "I was born with carving" (Fischer, "Künstler," 1963, 213). Momo Fina told me in 1980 that "the only way I pick up the carving—I'm born with it. It is gift of God. I was born with it, nobody taught me." And Fischer and Himmelheber quote another Dan carver, Tame, who describes his visualization before carving, "All the masks which have ever been made by men ... pass before my mind's eye like something which floats past on the surface of the river" (Fischer and Himmelheber 1984, 188).

Even if he has been born with the gift of carving, and perhaps has had a lengthy apprenticeship, the actual process of carving is one that the artist views as difficult. As Momo Fina told me, "To come to be a carver, it takes you a long time. For you to take your hand, pen, so ... to draw a mark and carve human being, his nose, his eyes— everything correctly, it took long time. And for you to carve something, carve something that finished will look like a mask—dancing mask—it took a long time. It's very hard-o! Carving is very critical work."

Thus, several commonly held attitudes seem to exist among the carvers in Liberia. What is interesting and important about the research on the Dan carvers is that three successive generations of carvers are known, providing historical depth that is rare in the study of African artists. Because of this we are able to see change as it affects style and technique. Also, Barbara Johnson has been able to follow up an obscure reference from Himmelheber concerning one of Zlan's wives who possibly carved. As is described in her essay, Johnson found a niece of Zlan's who carves today. This information, once it is thoroughly documented, may help overturn one of the maxims of African studies, that women never carve.

There are several differing or overlapping ideas about how and why stylistic change occurs. The recognition that significant change does occur has been important in shattering the old notion of African culture as traditional and static. It generally is agreed that change in African art results from such external influences as trade, warfare, Islamic and European penetration, western education, medicine, and technology. Certainly, all of these are significant, as is the changing marketplace brought about by these influences. William Bascom points as well to another origin of change. "The only possible source of divergence and specialization of art styles which have derived from a common origin lies in innovation by individual artists" (Biebuyck 1969, 99). From this perspective we can acknowledge the artist, who is striving for recognition and fame, as a direct source of change, assisted in the case of artists in Liberia by their spirit helper. As the creative artist is exposed to new materials, technologies, and markets, one might expect change to be rapid. Because of the conservative nature of the social structure, rapid innovation is inhibited and stylistic change takes place gradually.

Connoisseurs generally have lamented stylistic change, as they think the aesthetic quality of African art has deteriorated over the last several decades. Bascom describes the phenomenon of giganticism. To appeal to western markets, objects are enlarged, made grotesque, decorative, or elaborate. Similarly, western art historians have interpreted the movement of art from classicism to periods of baroque and decadent art.

Varney Massalay, the Vai carver I worked with in 1980, carves primarily for a tourist market. As he says, "The people of ELWA Mission send visitors to me to make things. People will spend two or three months here. I'll do their work for them and they will carry it." Most of what he makes is art that has no formal or stylistic relationship to the art used by the Vai people, which he also occasionally makes. Varney can make almost anything requested, getting his sources from tourist clients. "All the kinds I can see I can copy them." The faster the carving is made, the better for the client with limited time who wants a souvenir, and thus better for the carver, who makes a sale. This rapid production is referred to by Momo Fina and Varney as "quick service." Thus, if stylistic quality has deteriorated, it might be attributed to the carver-client relationship that no longer is bound by the conventions of the carver's own society. However, artistic innovation may result, as in the case of Momo Fina, who "invented" chess sets for his western clients. These sets depict figures from tribal life, such as a very important masked figure as king, or a fowl as the pawn. These creations illustrate the positive side of a new, non-traditional, carver-client relationship.

However, the decline of artistic quality, because the artist either is working now for money or trying to get more money by producing rapidly, must be considered, even though the artists deny money is a factor. Nelson Graburn suggests that "in the headlong rush to please the tourists and tastemakers the artisan finds himself in danger of surrendering control of his product. Where this has occurred it is no longer his art, it is ours. He is now subject to our manipulations and our aesthetic whims" (1976, 32).

These changes can be seen in the work of the three generations of Dan carvers examined by Barbara Johnson. Zlan, the oldest of the three, is still revered by his people; those works attributable to him certainly are the most successful among the carvings examined in this exhibition in terms of their form, finish, and detail. He produced exclusively for the Dan society as did his pupil, Zon. Zon, however, seems to have developed one limited style and stuck to it almost slavishly. Although Zon was an excellent carver, he did not innovate and was not inspired by dreams or by other artists. His work, while proficient, lacks some of the power and depth of his master's. Dro, a relative and pupil of Zlan, seems to have been greatly affected by a changed marketplace and society. His primary market has been other than the Dan, although he sometimes carves objects for traditional use. His style, technique, and imagery seem less inspired, possibly because much of his production is directed to the tourist market. In fact, I recently learned that game boards by Dro are being sold by traders outside the supermarket in Liberia's capitol (William C. Siegmann, personal communication, 1985).

For the Dan carvers discussed in this catalogue and the Vai carvers I have worked with, it seems true that an explicit relationship exists between the quality of a carving and the client for whom it is made. Nevertheless, much more research must be done among living African artists in an effort to try to understand the very complex dynamic that exists between the artist, the larger national and international marketplace, the changing society, and the creative process. It is hoped that the artists who are to be studied will be more forthcoming than the Gola woodcarver who told Warren d'Azevedo, "Chameleon does not let anyone know his true color" (d'Azevedo 1966, 67).

Thomas K. Seligman
Deputy Director for Education and Exhibitions
Curator in Charge, Department of Africa,
Oceania, and the Americas

History, Social Organization, Politics, and Religion of the Dan People

The Dan, who number approximately 350,000, live in inland Liberia and the Ivory Coast. Their land ranges from forest in the south to a rolling savanna in the north. They are a farming people who annually clear forest land to grow the cash crops of cocoa, coffee, rubber, and their staple foods of rice, cassava, and sweet potatoes. Most protein is derived from fish in the streams and forest game; livestock is raised for special feast occasions. Greens are gathered from wild and domestic plants, vegetables are cultivated, and palm oil is extracted from the nuts of the wild oil palm.

History

The Dan, who speak a Mande language, originally came from the western Sudan region to the north, part of the present-day states of Guinea and Mali. People have moved from this area to the south since about the eighth century to the present day, forced by expanding political states and the continual advancement of the Sahara due to dessication. The location and movements of the Dan, Mano, and We can be reconstructed as early as about 1300 (Person 1961, 47–49). At this time the Dan and Mano (also Mande speakers) were located in the savanna region of the northern Ivory Coast and Guinea. The We, who speak a number of Kruan languages, were in the northeastern Ivory Coast. From about the first half of the sixteenth century, the Dan and Mano began to move in parallel lines to south of the Nimba range into the high forest, an area sparsely populated by the Bassa and We. The Bassa were pushed south, the We south and west (Person 1961, 52–54). This southern movement of the Mande speakers was influenced by two factors: political turmoil in the northern kingdoms, and population growth with resulting land depletion, causing a need for new land. These migrations were not national movements, but were made under the direction of "big men" who took followers with them and formed towns. Smaller groups followed to form new village segments (Riddell 1970, 27). By about 1700 the situation in this area had stabilized (Person 1961, 31). New towns were continuously formed when one segment of a town would break away under the direction of a leader, and move to a new territory nearby.

Proud of their fierce reputation in war, for many years the Dan fought constantly with their neighbors. With the establishment of the new independent nation of Liberia in 1847, which was governed by freed American blacks who had returned to Africa beginning in 1822, the new central government in the coastal capital city of Monrovia began to pacify the warring peoples of the interior. Pacification of the Dan region in earnest took place in the early 1900s in response to the threat of border encroachments from colonial governments to the north and east, and was largely completed by the time of the outbreak of World War I. At this time peace was established, administrative controls were set up by the central government, and movement and trade were facilitated.

Social Organization and Politics

The basic unit is the family, comprised of a husband with one or more wives and their children. Traditionally, each wife and her children inhabited a separate house, although today they commonly live in separate rooms in a single large house. Men marry women from other lineages, usually from other villages, and bring their brides to set up housekeeping near their own father's house. Inheritance and descent are patrilineal, traced from father to son through the father's side of the family. Lineages, or groups of people descended from a common ancestor in the paternal line, form the core of each of several sections of the town, which are called *quarters* in Liberian English. Each quarter selects as its spokesman a quarter chief, usually the eldest male, or sometimes the one with the most forceful personality. The town chief is selected by and answerable to the village elders; his main function is to mediate between the central government and the people. The real controlling power of the town, however, lies in the hands of the council of elders, and nothing is decided before first consulting them.

In Liberia the English term *clan* is applied not to a kinship group, but to a number of towns administratively grouped together by the central government. The clan is headed by a clan chief. Similarly, paramount chiefdoms are made up of a group of clans. Usually, paramount chiefs, chosen by the central government, were powerful chiefs in their pre-centralized culture. However, before central government control, towns functioned as separate entities and only occasionally formed temporary alliances for military purposes.

Men's societies, also controlled by the elders, form the real socio-political unit of power in the Dan community today, as they did in the past. Although the Dan do not have Poro, the intercommunity centralized men's secret society found among some Mano and other peoples of Liberia and Sierra Leone, most Dan towns have their own local men's societies. These societies demonstrate their power and effectiveness by calling upon and controlling tutelary spirits from the bush. These appear as masked figures in body-concealing costumes. Using these mask-spirits, the societies are able to settle disputes too great for the quarter or town chiefs, apply and enforce rules for the good of the community, teach and correct behavior, and validate the actions of the elders while demanding absolute loyalty and obedience from all society members.

All males are initiated into the society by attending *bon,* bush school, some time before or during adolescence. Boys are secluded in the bush (forest), away from the village and their families, to undergo certain rituals and be circumcised. This experience prepares them to encounter the mysteries of the spirit world and to learn the rules of adult men, above all, that of loyalty to the society.

Women of the Dan also have a society and bush school where they undergo clitoridectomy and initiation into adult womanhood. They wear no masks, however, and instead call upon tutelary spirits in other forms, including those of feast ladles, to provide them with the spirit help they need.

Religion

The Dan believe in one God, Zlan, who created the universe. Zlan created the earth and then, by shaping mud in his hands, formed people and animals. Zlan was, therefore, the first sculptor. Though Zlan is acknowledged, it is believed that mankind cannot reach him, and he receives no form of worship. Instead, there is an independent spiritual power that can be made to work for people because it desires to enter into the village life. This spiritual power, called *dü*, is appealed to for many kinds of help.

Dü, though invisible, is a force that is present in all aspects of the universe. It animates humans, leaving the body when a person dies and going to a "village of the dead" while awaiting reincarnation, usually into its former family. If the deceased's dü has been evil and has practised witchcraft to harm others, this dü will be banished by Zlan forever, never to be reborn. Dü from an ancestor can also manifest itself as an animal living in the sacred grove of the forest; that type of animal then is forbidden to the villagers as food. Dü also may be shared between a person and an individual animal. Then the person at times may be transformed into that animal, or may take on some of the animal's qualities when needed, such as great strength.

People harness dü to work for them by making an object or by designating a feature of nature for the dü to embody. Through such a visible, tangible form, people can communicate with the dü and offer sacrifices to it, in this way influencing it to act to their advantage. However they must wait until the dü announces its presence in a dream.

Dreaming is the vehicle by which dü communicates with people. It appears in a dream in the specific form it wishes to embody; for example, dü may wish to appear in the form of an animal horn that has been filled with specified substances known for their indestructibility or other qualities. Another form in which dü often manifests itself is miniature masks. To embody the most powerful form of dü, called *gle*, larger face masks appear first in dreams as part of complete masquerade ensembles. Besides making this communication to specify its desired earthly form, dü also conveys specific power and ability through dreams. Any person who has become great at something, or who has achieved success, is presumed to have acquired dü through dreaming, whether or not dü took a tangible form in the dream. Carvers interviewed in 1983 and 1986 nearly always mentioned dreams as instigating their carving careers. It was revealed to them in dreams, as well, how to carve and what to carve. Even when they acknowledged a master carver who taught them the technique of carving, greater emphasis was placed on the power and ability gained through dreaming. When the Dan oral historian Gbangor Gweh of Beple was interviewed about Ldamie the brass caster, whom he had known, and was asked how Ldamie had become famous, he replied that he did not know what special kind of power Ldamie had used to become famous, and could not tell us about Ldamie's dreams. Nevertheless he acknowledged that "there were and are people who can dream and thus make their activity famous" (23 March 1983).[1]

[1] For further discussion on these topics see Schwab 1947; Himmelheber and Himmelheber 1958; Fischer 1967; and Fischer and Himmelheber 1984.

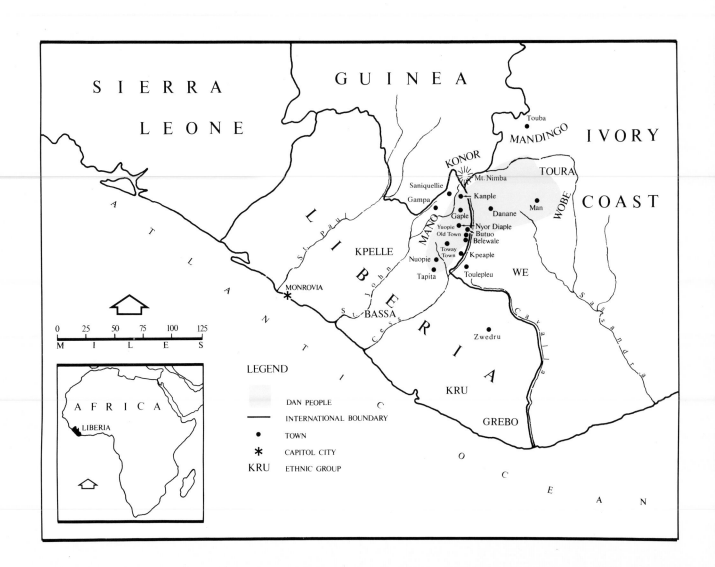

SIERRA
LEONE

GUINEA

IVORY

MANDINGO

Touba

KONOR

TOURA

Mt. Nimba

Saniquellie

COAST

Gampa

Kanple

MANO

Danane

Man

WOBE

Gaple

Yuopie
Old Town

Nyor Diaple
Butuo
Belewale

LIBERIA

KPELLE

Nuopie

Toway
Town

Kpeaple

WE

Tapita

Toulepleu

St. Paul

John

MONROVIA

Cavalla

BASSA

Cess

Zwedru

Sassandra

A
T
L
A
N
T
I
C

KRU

GREBO

O

C

E

A

N

0   25   50   75   100   125

M   I   L   E   S

AFRICA

LIBERIA

LEGEND

DAN PEOPLE

INTERNATIONAL BOUNDARY

TOWN

CAPITOL CITY

KRU   ETHNIC GROUP

Ritual, Prestige, and Decoration:
The Art Forms of the Dan

Masks

Long, long ago people made a law that the word of the mask be decisive.—Zlan, the carver (Himmelheber, *Negerkunst,* 1960, 143)

Masks are the most important art form of the Dan. Many of the other forms of sculpture are derived from the mask and what the mask symbolizes. Numerically, more masks are created than any other form of sculpture. Spiritually, masks are perceived to embody the most powerful of spirit forces. Socially, masks are the means of bringing control and order to village life. Masks provide the strongest impressions of a young Dan person's earliest experience, as their importance is reinforced by their presence at all significant events.

Masks are empowered by the strongest of supernatural spirit forces, called *gle.* Like dü, gle inhabit the dark forest, particularly where the trees grow high and dense. Gle long to enter into and participate in the ordered world of the village but, being invisible, cannot until a visible form for each is made. The nature of that form, a mask and complete masquerade ensemble meant to represent the personality of the gle, is seen in a dream. In addition, the gle must reveal its intended function in the dream or that dream is considered useless. The dreamer, who must be an initiated member of the men's society, reports the dream to the council of elders. They then decide whether the masquerade ensemble should be created for that man to wear and perform.

The carver carves the wooden face, and this is accompanied by attire that includes forest materials such as raffia, feathers, and fur. It is believed that each gle has its own personality, character, dance, speech patterns, likes, and dislikes, and it is given a personal name. The wearer of the mask takes on all these characteristics and qualities when he wears the mask ensemble. Having come from the unknown realm of the dark forest, a gle is thought to be unpredictable. Therefore it always has an attendant with it to control it as well as to interpret its speech.

Gle

Gle can be divided into two broad categories: that of *deangle,* gentle, peaceful gle, which have no gender, but whose qualities are thought of as feminine; and that of *bugle,* gun or war gle named after the sound of the gunshot (Tabmen 1971, 18), whose qualities are thought of as masculine. A third category, *gle va,* are "big" or "great" gle that have risen to importance, and can be either deangle or bugle in form.

Apart from these general statements, it is difficult to classify the many forms of mask gle to correlate form with function. The individuality of gle make this so. A gle in mask form, which might look very similar to another, could have different characteristics and therefore different functions, even in the same village. Furthermore, there might be

Fig. 2   Dancing *deangle* of *sohngle* type,
named Korto
Gaple, March 1983

Fig. 3   *Deangle*
Nyor Diaple, February 1986

Fig. 4   Woman with midline forehead tattoo
Near Tapita, February 1986

Fig. 5   Girl with kaolin face
decoration across eyes
Yuopie Old Town, February 1986

changes in function during one gle's lifespan, which is often several generations long. A new face mask could be carved to replace a worn out or damaged mask for the same gle. Confusion also enters with the many different names given to each gle. A gle is given a personal name (e.g., Slü, "Hawk"; Ble, "Termite"; or Korto, "You don't make farm," meaning the gle distracts one from normal work) and one or more praise names (e.g., Zuku, "Amazing"; or Sadhoplo, literally, the palm leaf funnel that enables one to pour palm wine, meaning that the gle enables one to achieve success). The gle also may be called various names that denote its functions or physical characteristics, or even names that distinguish it by the traditional implements it carries. One gle may have seven or eight names.

In spite of the possibility of such variations, the following range of known functions may be assigned to the most common mask forms of the Dan.

Deangle

The *deangle* mask-being has the oval face with slit eyes (figs. 2, 3; cat. nos. 1–4, 8, 17, 27). Its character is gentle and graceful, and it represents an idealized Dan form of beauty. Slit eyes, or eyes that barely show below lowered lids, are thought to be beautiful, as are the expressive protruding lips with a few teeth showing, the curve of the forehead, and the oval face. Deangle's function is to teach, entertain, and nurture; in other words it supports peaceful activities in the village. Although it has no gender, its qualities are thought feminine by both Dan and outsiders.

The deangle mask often has raised tattoo markings forming a vertical pattern or line in the center of the forehead, representing a discontinued practice of the Liberian Dan still seen on some older people (fig. 4). A horizontal band of white across its eyes is also often seen, representing the continuing cosmetic practice by women and girls of painting white kaolin on the face for special occasions (fig. 5). Among the northern Dan of the Ivory Coast and Guinea, deangle masks tend not to have the forehead line and instead sometimes have an incised pattern around the periphery of the mask, also representing tattooing.

Bonagle

The most frequently made and used mask in the deangle category is the *bonagle,* which serves the *bon,* or bush school. Every boy enters the bon at or near adolescence for one to four months, to be circumcised and initiated into the men's society. The bonagle acts as an intermediary between the bon and the village. It transmits messages back and forth, and carries food from the mothers to the boys.

Entertaining Deangle

Other deangle entertain and teach in the village. Singing deangle entertain and teach by singing accounts of history or memorable recent events; dancing deangle entertain and sometimes teach correct behavior. Deangle whose function it is to entertain tend to be dressed more elaborately than the bonagle, wearing richer garments and sometimes a more elaborately carved face mask. One type of gle in this category is the *sohngle,* named for its tall conical hat that looks like a basketry fish trap, *sohn.* The sohngle is said to be the best dancer, but it jokes, sings stories, walks with great dignity, mysteriously drinks water or palm wine with face mask in place, and imitates the birds of the forest as well (Kerser 1986, Tabmen 1986; see fig. 2).

Weplirkirgle

Weplirkirgle, the "fun-making" gle, is another form deangle may take (fig. 6). Because it sometimes has projecting tube eyes, weplirkirgle may also be a form of bugle. It is distinguished by its distorted asymmetrical features, usually representing a physical deformity. Although it is against the law to laugh at it, this gle is purposely very funny, thereby collecting fines. It makes fun of people with deformities, assuaging their feelings of being cast out, and alleviating the tensions that surround them. This gle teaches by negative behavior. Although we may safely assume that a mask with distorted features functions or functioned as a weplirkirgle, a mask with symmetrical features may also function this way by distorting its speech, coughing, or mimicking other deformities (Kerser 1983 and 1986; for other examples see Harley 1950, plate IX).

Bugle

The first mask to come to man in his dreams said, "Make me, wear me; that will scare the enemies."—Dro (Himmelheber, *Negerkunst,* 1960, 159)

   Bugle's eyes are usually either triangular or round, projecting as tubes, or carved out (cat. nos. 5–7, 18–20, 28). The face tends to be less flat and to have boldly projecting surfaces. It is meant to be fearsome because it functions to create excitement, "to make the town hot" (Kerser 1983). The personae of these gle once were associated with warriors, and whipped battle participants into action, frightened the enemy, and sometimes led the soldiers into battle. They dance with appropriate vigor. Bugle has been labeled masculine because of this aggressive behavior, although, like deangle, it has no gender.

Kagle

One functional category of bugle takes several forms and is called *kagle* (figs. 7, 8, 9; cat. nos. 6, 28). Kagle is the rough and vigorously dancing gle that takes its name from the hooked sticks it carries and hurls into the crowd. (*Ka* means "hook".) Kagle whips bystanders, pulls off shirts with its stick, throws the sticks indiscriminately into the crowd, and generally rouses the people (Tompieme 1983). It is believed that this was done in former times to excite the town in preparation for battle. Despite its rough display, it is greatly loved, particularly by men who make a game of avoiding its hooked stick and other forms of attack. Kagle is said to be associated with the chief, who used it to test loyalties during times of war. If the kagle injured anyone or caused damage with its rough behavior yet there were no complaints or criticism of the chief, that chief

Fig. 6 *Weplirkirgle* mask
Peabody Museum, Harvard University
37–77–50/2699

Fig. 7 *Kagle*
Form of *bugle*
Yuopie Old Town, February 1986

Fig. 8   Cubist *kagle* mask
The Metropolitan Museum of Art, New York
1979.206.219

Fig. 9   *Nya gbo kagle*
Nyor Diaple, March 1983

knew he could count on unfailing loyalty from the particular family involved (Chief Tomah of Butuo, in Thompson 1974, 164). Today kagle is followed by an attendant who carries the sticks and tries to prevent injuries.

All representations of kagle wear a wig of shaggy cotton yarn, have feathers stuck behind both ears, and wear a cloth cape and raffia skirt.

Cubist Kagle—The face mask of the kagle can be one of several types (Kerser 1983, Tompieme 1983). The cubist kagle has protruding disc-like or triangular cheekbones, triangular eye holes, an overhanging brow, a large open mouth, and no chin (fig. 8; cat. no. 6). Most of these masks are meant to look like chimpanzees and the wearer acts accordingly. A number of other masks offer variations on this same basic style, representing animals such as crocodiles or birds.

Nya Gbo Gle—Another form of kagle is that of *nya gbo gle,* which refers to its protruding, tube eyes (*nya,* "eyes," and *gbo,* "ceramic pot"; fig. 9). All masks with round or protruding eyes are called nya gbo gle, regardless of function. This type of kagle also has a prominent brow, often carved with a row of duiker horns. These represent power objects, as duiker horns are often used as containers for magical power substances. Instead of horns, the mask sometimes has a brow represented as a shelflike overhang. The jaw and mouth are either realistically human or large and animallike. These masks frequently combine human and animal features, again often resembling the chimpanzee.

Kagle with Slit Eyes—A third form of kagle features recessed slit eyes, broad mouths, and sharply projecting animal features (cat. no. 28). Its behavior is much the same as the others in the kagle category.

Blua Gle

Another functional type in the bugle category is the *blua gle.* This is considered an important gle whose principal function at one time was to escort and bless warriors. Today it investigates and settles disputes too great to be settled by the townspeople and often rises to the rank of *gle va,* the most important gle of an area, which is discussed below. Blua gle dances vigorously, showing its great strength by carrying someone on its back while dancing, or by picking up and flinging a heavy wooden mortar great distances (Kerser 1983).

Blua gle has a large jaw and projecting tube eyes, which make it by definition a nya gbo gle. The significant features that identify it, however, are the *blua,* the tall vertical black feather headdress it wears, and the *maan,* the broom whisk it carries in its hand (Tabmen 1986).

Gunyege and Zakpai Ge

Round-eyed masks, which occur only among the northern Dan, are another form of bugle. They are oval and have the fine features associated with deangle, except for their round eyes, which may be slightly projecting or carved out as holes. They are often painted red, at least in part. Functionally such masks fall into two categories, although the face masks themselves without their accoutrements often look much alike.

The two functional categories of northern Dan bugle masks are called *gunyege* and *zakpai ge*. The northern Dan equivalent to gle is *ge*. Gunyege participates in races with the fastest young men of the village. It is believed that the power of the ge helps its wearer to win; if the wearer of the mask loses the race, it means that the spirit has deserted that person, and the mask will go to the fastest runner. In this sense it is a trophy mask. The mask usually has wide-open round eyes, necessary for good vision, and is tied to the head with a strap. A kerchief usually serves as head covering (fig. 10). Zakpai ge is the fire prevention ge. Its function is to insure that women have put out their cooking fires every day during the dry season, before the afternoon winds begin to blow. Zakpai ge is aggressive, sometimes throws things, and is meant to inspire fear. The round eyes on this mask are often smaller than those of the gunyege, and may even be slightly projecting tubes. Tall green leaves cover the head. In addition, the masker wears pants with a ruff of raffia around the waist and neck. It carries a branch as a weapon (Fischer 1978, 21).

Ge Gon

Another mask type also exists only among the northern Dan and their neighbors. This mask is called *ge gon,* "masculine" ge (fig. 11). It appears to be a variation of bird-beaked masks from the Mau. The mask has oval or slit eyes often framed with tin, and a large beak or snout with a black beard of monkey fur. The lower jaw is often moveable. The headdress of this ge is ornamented with mirror glass, cowries, cloth, fur, and tall white feathers. As the mask spirit swoops and dances, the masker carries in its hands horsetail fly whisks that are waved gently to imitate a large and graceful bird (Fischer 1978, 22–23). The bird represented by the ge gon is probably the hornbill, important in Dan mythology as the first being created by Zlan and as bringer of the oil palm, which is an important food source of the Dan. Ge gon today dances strictly for entertainment, although it is thought long ago to have sung songs and proverbs to instruct the people in the importance of this mythological bird (Fischer and Himmelheber 1984, 81–85).

Tankirgle and Gbaagle

Most of the above gle come under the broad heading of *tankirgle,* "dancing" gle (literally "to make dance" gle; Tabmen 1971, 15). Another term for dancing gle is *gbaa gle,* literally "bench" gle, so called because the figures perform in a defined area where spectators can watch while seated on benches (Tabmen 1986). Although these gle are admired and respected for various manifestations of spiritual power, a skilled dancer is greatly loved also for its entertainment value.

Fig. 10   *Gunyege* mask
Private collection

Fig. 11   *Ge gon* mask
Peabody Museum, Harvard University
48–36–50/7317

Gle Va

Gle va, big or great gle, are the most important, most powerful gle of an area (cat. nos. 7, 9). In the past a gle va had an enormous reputation, and rarely made an appearance. It only settled large matters, such as stopping wars between villages. Some became so famous for their ability to negotiate settlements that they were requested across great distances and ethnic boundaries. Himmelheber tells of the gle va named Ve, who was Mano but was sent for by relatives in We country to stop a war between two We towns. Ve was known as the gle who could separate fighters (Himmelheber, *Negerkunst,* 1960, 146).

Today a gle va settles disputes that cannot be settled by ordinary authorities. It also maintains traditions by policing and controlling traditional ceremonial activities, such as the opening of bush school or a cow feast (Fischer 1978, 23).

Gle va may wear the mask of any deangle or bugle who has been promoted to this highest rank, although it is most likely to be embellished with fur, feathers, leopard skin, cast brass bells, and usually a red shawl. If a mask has been carved to be a judge mask, it can be large, with nya gbo, projecting tube eyes, and a moveable bearded animal jaw hung with symbolic power objects (Kerser 1986, Tabmen 1986).

Gunlagle and Wompomeingle

To rise to the status of the most important gle of an area, a gle must prove itself wise in settling disputes and powerful in bringing about desired results. It must prove that its spiritual backing is strong by demonstrating supernatural feats. *Gunlagle,* the "village quarter" gle, is the judge of its village quarter (section of village divided by family lineage). Disputes within the quarter are brought before, and settled by, the gunlagle; a good and wise gunlagle can become a judge for the whole town. A *wompomeingle,* an "accusation" gle, judges and settles disputes for a whole village or larger area (cat. nos. 18–20). It commands respect wherever it goes. One must be careful not to break any of its many rules of behavior in its presence for fear of being fined in such valuables as white chickens, kola nuts, or money (Kerser 1983 and 1986, et al.). Thus the problem in identifying these masks by function alone is very great when we consider that any gle, if it proves itself worthy, can rise hierarchically to become a leading mask, or gle va.

Gor Gle

The problem of identifying gle va becomes even more complicated when we consider the *Gor,* "Leopard" society, which moved into some regions of the Dan from the northeast in the last century (Fischer and Himmelheber 1984, 105). The principal function of the Gor seems to be the administration of justice, and as such it has superimposed itself on the existing system of men's societies that control the mask spirits. As it usually was high-level society members who joined the Gor, their mask spirits became elevated in status by being invested with Gor functions, and they often became leading masks, gle va. These masks would also be known as *Gor gle* (Fischer 1978, 23).

Miniature Masks

Every morning, in secret, the owner takes out his ma, *spits on its face, rubs its forehead against his own and says, "You there, good morning. Don't let any witch come to me. So be it."*
—George Schwab (1947, 365)

Miniature masks, *ma go,* like the big masks, are carved to embody tutelary spirits, dü (cat. no. 21). The main function of the ma go ("small head") is the protection of its owner from harm, particularly that caused by witchcraft. However, it also may be used in divination and as a sacred object upon which to swear an oath. Thus miniature masks are treated like other sacred objects, regularly fed with ritual offerings and kept hidden from public display. The forms of ma go vary almost as much as the forms of the masks meant to be worn, but, like the full-sized masks, the slit-eyed deangle type predominates. These small masks also vary greatly in artistic merit, as it is not considered difficult to carve one and many attempt it. Particularly well-carved and well-treated ones may be miniature works of art, having acquired a fine patina from being "fed," rubbed with oil, and carried around for a long while.

Personal protection is afforded by a ma go, which may take the form of a particular full-sized mask. In such cases the *zo,* or owner of a full-sized mask, as well as other male members of his family and occasionally even women, may keep and carry a miniature of the large mask to serve as a ma go (Fischer and Himmelheber 1984, 107). This allows the individual a personal and portable form that shares the power and protective force of the full-sized mask. Moreover, the person is provided with a symbol that he is under the power and protection of a particular mask. Such a symbol may occasionally be shown to demonstrate a person's status and affiliation in places where he is not known (Si 1983). This function may have given it the commonly applied name of "passport mask."

The ma go is appealed to in times of uncertainty and danger. Before undertaking a journey, for example, the owner rubs his ma go with palm oil and asks aloud for its protection (Schwab 1947, 278). If he is to appear before a court hearing and wishes to present a strong case, he will again appeal to his small mask. If the owner or one of his family falls ill, a native healer sometimes will advise commissioning a carving of a ma go in order to cure the illness by spiritual means.

Ma go also serve to protect members, collectively, at men's society meetings. Here ma go are laid out in a flat winnowing basket, along with other protective objects embodying dü. Sometimes a ma go is attached as a face to a cloth figure (Himmelheber, *Negerkunst,* 1960, 161–162); or it may be attached to an animal horn filled with power substances.[1]

Ma go also are used as sacred objects for taking oaths. Upon entering bush school, every boy or man must swear on them that he brings no harm. Likewise they are used for swearing to tell the truth at any hearing in the bush school (Schwab 1947, 277–278).

---

[1] There are a number of these in the Koninklijk Museum voor Midden Afrika in Tervuren, Belgium.

Fig. 12 Dan man
with *ma go*

Native healers use ma go divination to help them arrive at a diagnosis. For example, "One breaks a kola nut in two parts, presses them on the forehead of the mask, and spits on them. Then one throws the halves of kola on the floor." The prediction is based on the pattern made by the nuts falling cut-side-up or down (native healer, in Himmelheber, *Negerkunst*, 1960, 162).

A ma go, as are all other objects that are kept to embody dü, must be fed regularly to keep it strong and able to help its owner. Food may be simply set before it, or the offering, such as rice or oil, may be rubbed or poured onto it. Kola nuts may be chewed, then the juice spat onto it; on special occasions a sacrificed chicken's blood is spilled onto the mask. Unless regularly fed in this fashion, the spirit, the dü, grows weak and tired from working on behalf of its owner and can be effective no longer (Schwab 1947, 364).

Feast Ladles

The old woman promised me before she died that if I held the ladle in my hand, my name would become famous and the ladle would make me rich because it is so finely carved that, out of delight, the people would give me gifts.—Doa, a wunkirle (Himmelheber and Tabmen 1965, 179)

The *wunkirmian* or *wake mia* is a large ceremonial ladle (cat. nos. 14–16, 23, 29). Its name translates literally as "spoon associated with feasts." Such ladles are carved to honor a particular woman who has distinguished herself among her fellow women by generosity and hospitality. Such a woman is known as a *wunkirle,* or *wake de,* an "at feasts acting woman"; *mian* or *mia* is the name for any large ladle (Fischer and Himmelheber 1984, 123).

The wunkirmian is owned by the wunkirle, who is considered the "most hospitable woman" of her village quarter. One woman in each quarter is honored with the title of wunkirle (plural: *wunkirlone* or *wunkade*), an honor often handed down with the wunkirmian. When a wunkirle becomes old she chooses her successor from among the young women of her quarter. As the Dan are a patrilocal society, and wives usually come from other villages, she cannot pass on the ladle to her daughter, but instead chooses the woman she believes to be most generous and industrious to succeed her.

Himmelheber, with Tabmen, has summarized best the duties attached to the office of wunkirle (1965). She offers hospitality to all who come to her door, nor is any group too large for her to feed. Itinerant bands of musicians and entertainers, for example, find their way to her, and she traditionally delights in feeding them. In addition, she prepares meals for the men who clear the fields at planting time, and at festivals provides hospitality to arriving strangers. In order to be able to afford this largess, she must be an industrious farmer. She must have a husband or son who will do the heavy work of clearing large fields each year, and she herself must work long and hard to plant and harvest an abundance of rice.

At feast times she marches with her spoon at the head of the line of women from her quarter. Each woman carries a pot of cooked rice and soup. The wunkirle either distributes the food to the guests, or more frequently uses her ladle to indicate the distribution. At some feasts the wunkirlone of a village compete with each other in generosity by distributing small gifts of peanuts, candy, coins, and other foods. The women dance at these times. The wunkirle's prestige may be indicated by her being carried in a hammock through the village by the women of her quarter. They also contribute gifts of their own, but always in the name of their wunkirle. Guests in the village, for the sake of impartiality, decide which wunkirle is the richest and most generous, after which the masks of her village sing her praises (Himmelheber and Tabmen 1965, 174).

A similar tradition was described in some detail from several towns of the Dan-We border area (Kpor 1986, and Dro 1986). One woman from each town quarter is chosen as being the best cook, the strongest and hardest worker, and most willing to offer

Fig. 13 *Wunkirle* procession

hospitality. Every year or so on the occasion of a big cow feast, a particular quarter is designated to provide the cow. That quarter is also honored by providing its candidate to be the *klaywaiyno*. Early in the morning of the feast, after the cow has been killed and some of the meat cooked, the chosen klaywaiyno and her two women assistants dress in short, knee-length *lapas*, measures of cloth wrapped and tied around the waist, and brassieres. They paint white kaolin clay lines on wrists, elbows, ankles, knees, and across the eyes. The klaywaiyno also drags rice plants behind her from a long rope tied around her waist. She carries her badge of honor, the feast ladle, called *klaywaimina*, which corresponds exactly in form to the Dan wunkirmian. Her two assistants carry either ceremonial wooden pestles, sometimes with a carved head on the end, representing another function of women, the pounding of rice and other foods in the mortar with the pestle. In some towns they carry only rice plants.

The klaywaiyno and two assistants dance through the town, the klaywaiyno carrying her feast ladle filled with cooked rice, spread with red palm oil and cooked meat, stopping as she goes to serve small portions to the most important men. Later the portions of the cow will be distributed to be cooked and eaten by all.

Among the Dan, after an old wunkirle dies a festival usually is given to honor her and to inaugurate her successor. At this time the new wunkirle must prove herself worthy of the honor; the ladle's function is reinforced as she occasions a sizable distribution of foods and gifts to her village and guests.

Besides being emblems of honor, these ladles have spiritual power. In the words of a wunkirle, Doa, the ladles contain "all the power and fame of the wunkirle." Ladles embody dü, and it is that power that enables the wunkirle to perform her duties in such a way that will make her rich and famous. Wunkirmian are the women's chief liaison with the power of the spirit world, and the authority symbol of that connection. The Dan say, "The wunkirmian is for the women what the masks are for the men" (Himmelheber and Tabmen 1965, 177). As are masks, each wunkirmian is given an individual name, e.g. Piase, "Fine face," or Mlanyor, "Lucky Woman" (interviews 1986). When a new wunkirmian is carved to replace an old one, the dü must be induced to enter it, and sacrifices are made to this end. After funerals, women dance in procession behind wunkirlone with their ladles, using the spirit power to chase away ghosts of the dead (Donner 1940, 88). Himmelheber tells of a Dan wunkirle whose ram-headed ladle had a horn broken off. According to the wunkirle, the horn was broken by fighting with another ladle in a meeting of wunkirmian spirits in the bush (*Negerkunst*, 1960, 166).

These ladles have a spiritual connection with the masks. Thus the wunkirlone, ladles in hand, often appear with the mask-wearers, tossing rice in front of them as a blessing (Himmelheber, "Sculptors," 1964, 245). When the gunlagle, the judge-mask of the village quarter, makes its appearance, all the wunkirlone of the village must participate, dance, and honor the mask by giving it gifts (Himmelheber and Tabmen 1965, 175).

Wunkirmian are carved in several different forms. The most common form has a handle carved in the likeness of a human head. The face is usually the oval, slit-eyed face of the deangle mask. It often has a vertical line from forehead to nose that represents tattoo markings, and a band of white kaolin frequently is painted across the eyes, representing a cosmetic practice of Dan women. The head usually has a carved hairdo, often trimmed with black-dyed plant fiber, sometimes representing hairstyles no longer seen but fashionable at one time. Decorative incisions often are made around the neck at the base of the head, and vertical lines are carved in eyebrows to represent cosmetic plucking. The neck often has incised rings, representing neck creases, which emphasize the beauty of a long neck.

Apparently the face on the ladle is intended to portray a specific woman. Harley writes of a Janus-faced ladle, said to have been carved about 1860 in We country to portray a wunkirle named Ma Boa, or "What thing do I lack?". On the back of the handle was carved the portrait of her favorite helper in feast preparation, Nying Gli, "Dry your tears" (1950, 40). Himmelheber and Tabmen also wrote and amply illustrated that the face on the ladle was indeed intended as a portrait of its original owner (1965, 176). A portrait in this context is a stylized one, emphasizing individual elements such as a special tattoo, scarification markings, or coiffures, yet staying within the traditional style.

Fig. 14   Feast ladle with legs
Collection Todd Robert Baker
Los Angeles

Another form that may be given to the handle of the wunkirmian is a pair of legs (fig. 14). This makes the bowl of the ladle represent the upper part of the body, presenting a wonderful abstraction of the human form. Legs on a feast ladle are said to represent the legs of all the people arriving on foot to be fed by the wunkirle (Woto Mongru 1983).

Other handle forms include the human hand, said to represent the strong grip of the wunkirle (Woto Mongru 1983). Animal heads also are seen, specifically sheep, goats, or cows, either representing the sacrificial animals at a feast or perhaps a woman's dowry (Zerlee 1983). Sometimes the handle ends in a little bowl, again suggesting food, or in a number of abstract designs.

Guardian Heads

Carved wooden heads on sticks, pedestals, or staffs are another class of objects found among the Dan. The pointed stick may be stuck in the ground, the pedestal allows the head to stand upright, and the staff may be carried in the hand. The heads and their attachments vary in size and form. They may be very small to nearly as large as the size of a mask. The most common face is oval and slit-eyed, resembling a deangle mask, or occasionally one with tube eyes resembles a nya gbo gle mask. The head usually is carved in the round; sometimes it is Janus-faced with two different faces represented (fig 15; cat. no. 22). Occasionally the heads are portrayed as two masks, back to back (fig. 16).

Not enough is known about the functions of these heads to relate form to function, or even to know whether there is a correlation. We do know that these heads are valued as powerful spiritual objects, the receptacles for tutelary spirits, dü. When their faces are carved in the likenesses of masks, they apparently take on some of the power of the mask spirits, gle, and the importance associated with the gle in Dan culture.

These heads may be carved to offer housing to spirits of the honored dead. Himmelheber tells of a meeting of elders that had to be delayed until a particular carved head representing a deceased elder could be fetched from a nearby farm and placed among them (*Negerkunst*, 1960, 168). Some are said to be portrait carvings of a living or recently deceased family member; after the person has died the guardian head is regarded as the embodiment of that person and is regularly offered food (Tabmen 1986). The carver Dro from Kpeaple told that these heads traditionally are given to babies because they represent the spirits of "persons who lived before" and who have returned. The baby keeps the carved head and when old enough might take that mask that the head represents (Dro 1983).

Donner writes of these heads as protective, guardian objects kept in the home, which occasionally are given offerings of palm oil and animal blood. Donner and Himmelheber both mention Norle, which is the name of the guardian head regarded as the spirit mother of the Snake Society. It is kept with all the society's other sacred objects which contain dü (Donner 1940, 85–86; Himmelheber, *Negerkunst*, 1960, 168).

Janus-faced heads on a staff are reported to be carried by certain masked performers. The two faces looking in opposite directions are said to symbolize the supernatural ability of the gle to see in all directions at once (Himmelheber, *Negerkunst*, 1960, 168). Another explanation for the two different faces on a Janus-faced head was given by the carver Dro, who is from a Dan-We border town near the Ivory Coast. He said that mask spirits, like humans, take spouses, and the two faces represent a spirit husband and wife pair (Dro 1983).

Agricultural societies such as the Bush Hog and Elephant Societies of the Dan and Mano, not to be confused with the politically significant village men's society, use the head on a pedestal or pointed stick called *kedie* ("clear ahead"; Fischer and Himmelheber 1984, 119). A kedie is placed in the ground or stood upright at the far end of the field to be cleared, and the men work toward it. Though they are all members of the

Fig. 15 Janus-faced guardian head
Wood, 25 cm
Buffalo Museum of Science
C12815. Acquired in 1938

Fig. 16 Janus-faced guardian head
portrayed as two masks
Stone, 7.2 cm
Collection William C. Siegmann

same society, each is in competition with the other to be known as the strongest and fastest worker. The kedie is perceived to give them increased strength and protection from accidents, believed caused through witchcraft by jealous outsiders. The first worker to reach the kedie at the far end of the field then has the honor of carrying it back to town at night, dancing and accompanied by music (Siegmann 1977, 35).

Figures

Most figures are pure art works, fashioned for the sake of creation itself.—Etta Becker Donner (1940, 80)

Although there are comparatively fewer figures than there are other Dan sculptural forms; *lime* ("wooden human beings") are an important form of art from this area. Dan carvers consider the lime a difficult form to make, the carver Zon stating, "It is most difficult to carve a figure, and next to that to carve a gameboard" (Fischer, "Künstler," 1963, 75). Besides the difficulty of carving, these figures traditionally have had a very limited market.

Usually formed from a single piece of wood, Dan figures generally are carved symmetrically, with frontal emphasis, and in an upright standing posture (fig. 17; cat. nos. 10–13, 24–25). The oval face is similar to the face of the deangle mask, having narrow slit eyes and expressive out-turned lips. The lips usually show several upper teeth of metal, sometimes set at a slant, in order to represent front teeth filed to a point, once a practice in this region (Donner 1940, 83). The neck may exhibit horizontal rings, representing creases, as also seen on some feast ladles. The rigid arms and legs are usually short in proportion to the head and body. Frequently, scarification patterns are seen on chest and abdomen. The coiffured hair, carved from the wood, sometimes has plaited fiber tresses attached. The lime usually is given a little loin cloth and some jewelry, such as ankle or knee rings, waist ornaments, necklaces, bracelets, and/or earrings.

The figures usually represent women, although sometimes men are portrayed, and then frequently in male and female pairs (cat. nos. 12, 13, 24, 25). Although they are stylized in the traditional manner, aspects of portraiture may exist in cicatrization patterns, coiffures, and other individual conditions such as the shape of breasts, umbilical hernias, or pregnancy (Donner 1939, 150). If meant as a portrait, the figure carries the name of the person it portrays (Donner 1940, 80–81).

Whatever their origins, during the past fifty years the figures became a secular status symbol. Apparently few were carved before that time. Donner writes of figures carved to portray the wives of chiefs, perhaps wives who had died and thus would be remembered, or sometimes a living wife. Upon the acceptance of a newly carved figure, the chief was expected to provide a big feast for the whole village, and to pay the carver as well. Thus a figure was an expensive object, one that only a wealthy man could afford, as it forced the display of wealth and its sharing. The owner of the figure then put away this status symbol, perhaps in a basket, and may have charged a small fee to show it to others (Donner 1939, 150).

The figure may have been commissioned, or it may have been carved at the instigation of the carver himself. He took the wrapped and covered carving upon completion to the chief, asking—sometimes even before showing it—what price the chief would pay (Donner 1940, 81). Even when the piece was commissioned, the price usually was not agreed upon in advance. Zlan mentions several examples of what he was paid:

I went to a chief of the Dan, named Bu, in Tuba . . . for him I carved a figure with a child on the back. For it he gave me a sheep, a goat, and a large garment. On Bu's command I carved the woman as his head wife looked. . . . I went to a place called Deu. There lived a rich man named Ba; he had called me. "You must carve for me a woman and a man, and four big rice bowls for my women." I made him even ten! In addition a stool. . . . Bu gave me a big cow, and slaughtered a second in my honor (Himmelheber, *Negerkunst*, 1960, 174).

The figures at one time may have had a religious meaning. Donner in the 1930s was told about the "long ago" practice of a ritual, held sporadically at the time of a new moon (1940, 81–82). There would be great feasting, dancing, and music, and people would travel from nearby villages to participate in the festival. The festivities seemed to center around a ritual of annointing the figures with palm oil and decorating them about the eyes with kaolin. Apparently both practices were ways of showing respect and honoring the figures, but the precise meaning of the festival and the significance of the figures has been lost.

"Every representation of the human face means contact with spirits and magical powers," Donner notes (1940, 79–80). All objects portraying a human face at one time were treated with caution and care. Sacrifices occasionally had to be made to them in order to appease the potentially dangerous spirits involved. As it was being carved, the figure was kept hidden from the eyes of the curious. Donner further states that *when a statue is only made in memory of some person and even when it concerns a "portrait" of a still-living person, there clings to the representation of a human face something disturbing and mysterious, and each piece of carving which contains the representation of a head is handled with a certain timidity* (Donner 1940, 80).

The Dan carver Zon (1983) stated that he only began carving figures after the introduction of "western civilized ways." He added, "Before that it was forbidden to carve the face. Secret society business was too strong." Although masks representing the face were carved long before the western world made its influence felt among the Dan, Zon may have meant that carving the human face for any other purpose than for the rituals of the secret societies was strictly prohibited. He thus implied again that the spiritual power associated with the representation of the face could only be dealt with effectively by the men's society, because of its experience in controlling masks and the spirit world.

Whatever the function of lime in former times, they no longer seem to have a functional role in Dan culture. Figures apparently have all but disappeared in the villages, having been sold off for the international art market. Unlike masks they have not been replaced. The only figurative carving done today appears to be for the export market, rather than for domestic consumption.

Fig. 17   Female figure
The Fine Arts Museums of San Francisco
71.26

Gameboards

Gameboards are another important sculptural form from the Dan carving tradition. They are carved for the game commonly known as *mancala* or *owari,* but among the Dan and Mano called *ma kpon* or *ma.* This game is known universally throughout Africa and southeast Asia. An ancient game, it was played in pharaonic Egypt, and is still popular among the Dan.

Carved from wood with a great deal of care, gameboards were and still are made to be handed down in the family (Schwab 1947, 158). They are considered among the most prized possessions of their owners. One of the reasons for their worth, according to the late carver Zon, is that they are an attraction to keep the children home and the family together, and to bring people into the home.

The boards are incised along the sides with linear geometric designs or depictions of lizards, snakes, or birds, the two sides having different motifs. There are twelve hollowed depressions, six on each side of the central divider. The moveable game pieces may be pebbles, pieces of metal, nuts, or seeds, but it is the fruit of the *ma* vine—a gray, smooth, and shiny fruit the size of a marble—that is most frequently used (Fischer and Himmelheber 1984, 138). Donner mentions one impressive set of game pieces that were cast iron in the shape of tree fruits (Donner 1940, 87).

The ends may be carved to form two cups for holding the game pieces. A carved human or ram's head, or occasionally both, may substitute for the cups (cat. nos. 26 and 30). The meaning of these carved depictions on the end of gameboards is unclear. Most carvers interviewed could give no reason for them other than that they were to make the gameboard "look fine." Their concern was with making the traditional designs look better than those of other carvers. All of the carvers interviewed, however, said that a human head could be a portrait that had been requested, possibly of the owner himself or his favorite wife. When not specifically commissioned in this way, a carver often will portray the remembered face of a beautiful man or woman (Zon 1983). Occasionally the face is carved without head and coiffure. In this case it represents the mask and the power associated with it. One old paramount chief, Woto Mongru, remembered a gameboard with carved masklike face commissioned and presented to a paramount chief by some of his people. This gift was intended as a warning to the chief that his power was limited and under the control of the spirit forces behind the mask (Woto Mongru 1983).

As for the use of the sheep or ram's head, the ram is thought of as a strong fighting animal. Warriors in earlier times associated themselves with it. Thus a master of the game enjoys his own association with the great strength and fighting ability that the ram symbolizes (Dro 1983).

With only minor variations, the rules are the same everywhere for the game. Two players sit facing each other. Allocated the row of six cups nearest to him, each player places four seeds in each of the cups, and then redistributes nine seeds, three from each of three cups on his side, "building forts." Players then take turns removing all the seeds from one cup and distributing them one by one in the cups, moving around the board

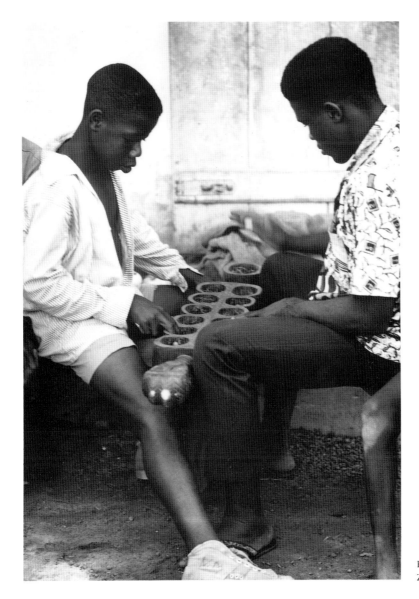

Fig. 18   Playing *ma kpon*
Zoulay, Liberia

in a counter-clockwise direction. The object of the game is to capture as many seeds as
possible. Capture is possible only if a player's last seed lands in an unprotected cup on
his opponent's side; that is, if the seed lands in a cup containing only one or two seeds.
These one or two seeds would be protected, however, if they lay behind a "fort," a cup
containing four or more seeds; neither could the one or two seeds be taken if more
than two seeds were in the cup or fort. The strategy of the game involves the even
spacing of forts and quick mathematical planning. According to de la Rue (1930,
183–184), the game is based on military strategy. The seeds are thought to represent
men, and the cups, villages. The object of the game is to "eat up" the "villages" of one's
opponent by capturing his "men." To play well, one must be able to remember the
number of men in each village, numbers that change with each rapid move. The game is
won when a player captures the greater number of seeds after one side of the board has
been cleared of seeds, or when one player is unable to make a move.

Fig. 19   Dan woman with brass jewelry in 1929–1930
See Schwab 1947, fig. 44

Brass Casting

For an apparently brief period, the casting of brass figures was an important art form in the Dan region. Typically, the figures stand about 8 inches high (20 centimeters) and either exhibit objects of importance in the culture or are genre figures that depict various activities in village life. (These figures are discussed in more detail in "Ldamie, Figurative Brass-Caster" in this catalogue.) Animals such as elephants (cat. no. 45), crocodiles, and dogs also are depicted. Changes in the times can be inferred from the number of brasses representing uniformed Liberian soldiers carrying guns and men in European-style hats.

The figures appear to have been made principally as prestige items and to have functioned in much the same way as the lime, or wooden figures. Like them, brass figures were either set out for display or carefully wrapped in cloth and stored in baskets, to be brought out and displayed for important visitors. At first only chiefs and wealthy men could afford this nonessential novelty, this "something different" (Yanemie Ldamie 1983).

In addition to brass figures, other kinds of brass castings deserve mention. A small number of ma go, miniature masks, are known in brass; these are usually of the highest quality (cat. no. 31). Ceremonial weapons and staffs were cast in brass as were everyday objects, including a chair reported to be in the house of Chief Mongru (Donner 1940, 52) and even a pair of sandals (Fischer and Himmelheber 1984, 156–159).

By far the most common form of brass casting, however, was the making of jewelry. Brass jewelry included necklaces that incorporated brass, glass beads, and brass leopard teeth (cat. no. 33), rings worn just below the knee and around the ankles (cat. nos. 34, 36), and bracelets. Sometimes these were worn in a series of graduated sizes proportioned to fit an arm or a leg. The jewelry was decorated with geometric designs, created by adding twisted, braided, or spiraled wax threads to the wax model before it was cast. Little brass bells often were worked into the jewelry design.

Both men and women wore brass jewelry, although women tended to be more heavily laden (figs. 19, 20). Schwab described the Gio (a frequently used name for the Dan of Liberia) in 1928 as being

aglitter from head to foot. Theirs seems to be an exuberant spirit, bursting forth in tinkling bells and clanging brass. With neck, arms, legs, fingers, toes so loaded with massive brass and iron ornaments that they can hardly walk with the weight of them; a grand head woman in Gio presents a dazzling spectacle (Schwab 1947, 112).

In the past men had worn brass anklets, but by the 1930s they were worn almost exclusively by women. Women's anklets usually were worn in pairs and might weigh eight pounds or more apiece. Anklets usually were attached at the time a woman was married and never removed until her death (Donner 1940, 55). The weight and richness of a woman's anklets and other brass jewelry reflected her husband's wealth, implying that the woman, who was considerably immobilized by the weight of the brass, was able to live a life of leisure while her husband's other wives or retainers did the necessary agricultural work and daily chores.

Cast brass jewelry was a widespread tradition, employing the technique of *cire perdue,* or lost wax, and providing the artists with the skills needed to cast brass figures. Some brass jewelry gave the wearer magical protection (Donner 1940, 53), but most jewelry was worn simply for prestige and adornment.

Copper and its alloys of brass and bronze historically have been the most precious metals in the traditional African system of values (Herbert 1984). For many centuries before the maritime trade to the West Coast of Africa, brass was imported in volume from across the Sahara into many parts of the continent. Undoubtedly, brass found its way in trade into the forest region of the Windward Coast, as the first European traders to arrive on this coast found a demand for copper and brass in specific forms (Herbert 1984, 120). For centuries European traders supplied this area with brass bars, *manillas* (bracelets used for currency), basins, and jugs (Johnston 1906, 74–75).

We know little about the antiquity of brass casting among the Dan as we have no documentary or archaeological evidence to demonstrate the age of the tradition in this region. Schwab (1947, 146) suggests that brass working may have come to the Dan from their eastern neighbors, the We, but this is not confirmed. What is clear is that the Dan had developed the art of casting brass jewelry and other objects by the nineteenth century and possibly much earlier.

By the end of the nineteenth century new sources of precious brass became available in the Dan area. The imposition of colonial rule in neighboring Ivory Coast and Guinea brought the arrival of a brisk trade in brass basins and other brass, including old European cannon barrels (Himmelheber, *Gelbgussringe,* 1964, 13). Moreover, the "pacification" of the Liberian interior in the early years of the twentieth century provided a new source of brass in the form of spent cartridge shells (William C. Siegmann, personal communication, March 1984). After peace was established, the trade of brass into the Dan area was made easier. For a period brass jewelry was made and worn in abundance.

By the late 1930s the use of brass jewelry declined. This may have been the result of a change in fashion, but certainly was also the result of a Liberian government decree outlawing all brass jewelry. The government maintained that the heavy anklets made the wearers prone to injuries and chafing infections. Possibly the government also felt that the heavy brass hindered the women from taking an active part in the development of agriculture and commerce in the interior of Liberia (Fischer 1967, 357). With the diminished use of brass in jewelry, an abundance of the material was available to be melted down and recast in new forms. This may have caused the development or at least the proliferation of the casting of full figures in brass (Siegmann 1977, 49). Very few brass figures show evidence of any great antiquity. The number of figures made by a relatively small number of individual casters, combined with the frequent representation of datable motifs such as Liberian military uniforms, helps assign this tradition mainly to the first half of the twentieth century.

There was no more than a handful of figurative brass-casters in the first half of the twentieth century. Donner described and illustrated a number of pieces (1940, 47–51),

Fig. 20   Dan woman with anklets
and knee rings of brass, in 1929–1930
See Schwab 1947, fig. 44.

claiming that there were only two figurative brass-casters in northeast Liberia—one Dan from Gaple (see "Ldamie") and one Kran or We from the Dan-We border town of Belewale. Himmelheber cited examples of the figurative work of two old casters (with Himmelheber 1958, 69, and *Negerkunst*, 1960, 184), and Fischer photographed a third artist in the process of casting a figure (1965, 93–115). The latter artist had not practiced the technique in years and did so only at Fischer's request. All of these casters are now dead, and with them, the art form.

The brass-caster was a much-respected member of his community (Donner 1940, 47). Brass casting was a specialty in its own right, but brass-casters usually were black-smiths, as blacksmiths were accustomed to working with metal and had the necessary tools. The blacksmith in a Dan village has always had a high social standing and has been well treated, because the community is dependent on him for its metal agricultural tools and weapons. Traditionally, a blacksmith captured in war was released if he could prove his metal-forging skill, a reciprocal, beneficial arrangement that avoided the loss of any practitioner of this essential skill (Schwab 1947, 145).

Unlike the practice in parts of West Africa of keeping the blacksmith physically and socially separate from the rest of the village, the blacksmith of the Dan village plays a central role in its political and ritual life. He belongs to the various cults, because his skill is needed for making cult objects. He is kept informed of political happenings, often performs certain ritual practices such as circumcision, and frequently becomes a high-ranking member of the men's society (Schwab 1947, 145; Holas 1968, 68; Fischer and Himmelheber 1984, 104). His shop is considered a favorite gathering place for the men of the community. As a sacred place where ritual objects often are kept, the blacksmith's shop is considered off limits to women and the uninitiated. In addition to their smithing skills, blacksmiths among the Dan frequently are wood-carvers as well; the need to carve wooden handles to fit iron tools often leads to further and more complex carving. Frequently, then, those who learned the specialized techniques of brass casting already were blacksmiths and wood-carvers, performing essential valued functions in the community.

The brass-casting process took place at night. Very few people were allowed to watch as it was thought that onlookers might steal the brass-caster's secrets and spoil his magic (Donner 1940, 48). The brass-caster had to observe certain rules and taboos. He must purify his heart and mind by banishing any evil thoughts and feelings, he must avoid breaking anything, he must abstain from sexual intercourse for the week or so required to produce a brass figure, and avoid all women in general because of their perceived general tendency towards witchcraft (Fischer and Himmelheber 1984, 144, based on their observations of two old brass-casters in the 1950s).

The Cire-Perdue Method

A brass-caster began with a lump of beeswax that had been softened by the sun. Pieces of wax were then kneaded, some rolled into threads and wound into spirals, others formed into the parts of the figure or animal being made. Later the pieces were assembled. Details were impressed into the wax, or applied with threads of lightly warmed wax. Sprues, conduits for the eventual pouring out of the wax and the pouring in of the molten brass, were added in the form of wax threads, often to the soles of the feet.

When the wax model was finished, it was oiled and then covered with a very fine, soft clay made from an appropriate soil, such as that of a termite mound (Gbangor 1983), or the loamy soil of an old house mixed with the juice of oil palm leaves (Himmelheber, *Negerkunst*, 1960, 184). It was important to mix sufficient carboniferous matter with the clay in order to absorb unwanted gases (Herbert 1984, 84). The most effective of these ingredients were the carefully guarded secrets of different casters. This first layer of clay was thin enough that the shape of the wax model could be recognized. A funnel opening was made to cover the wax sprues, and the clay allowed to dry slowly in the shade for several days. When this layer was well dried, other thicker, coarser layers of clay were applied and then allowed to dry. These outer layers often had animal hairs or pieces of bark added to make the investment stronger (Donner 1940, 48).

In small pieces, the brass was melted in a crucible placed in a hole in the ground. The hole had been surrounded by burning coals brought to a high temperature by bellows. The mold also was heated, the wax then either poured out and saved, or allowed to burn up and vaporize (Donner 1940, 48). The molten brass was poured into the funnel opening, filling in the shape formed in the clay by the wax.

The mold then would be immediately placed in cold water, hardening the brass (Gbangor 1983). In a few minutes the mold could be broken open and the brass piece removed. The piece then would be soaked in a concoction of cola leaves, followed by further cleaning with citrus juice to remove any copper residue brought to the surface by reaction to the heat (Fischer and Himmelheber 1984, 145). The sprues would be filed off along with any irregularities, and sometimes the surface would be hammered. Finally the brass sculpture would be burnished.

Fig. 21 Zlan
See Himmelheber, *Negerkunst*, 1960, 172
Fischer and Himmelheber 1984, 170

Four Sculptors—Three Generations

Zlan, Great Master and Teacher

I make everything especially well. When it should be a woman's figure I make her with a child on her back or a pot on her head.—Zlan (Himmelheber, *Negerkunst,* 1960, 173)

Zlan[1] was one of the great master carvers of this region; his influence was felt in Dan, Mano, and We towns in Liberia and the Ivory Coast. He carved for many wealthy men and chiefs, teaching many pupils from both the Dan and We peoples. Though he died over thirty years ago, he built such a reputation for himself that he is still widely remembered among Dan carvers in Liberia.

Zlan's carvings also are greatly admired as representative of the finest from this region, and are exhibited in museums and private collections in the United States and Europe. Museums such as The Metropolitan Museum of Art in New York, the Peabody Museum of Harvard University, the New Orleans Museum of Art, and the Musée National des Arts Africains et Océaniens in Paris have, until recently, treasured their sculpture by Zlan without knowing his name.

Shortly before he died after a long illness, Zlan was interviewed by Himmelheber (*Negerkunst,* 1960, 171). Zlan introduced himself: "My name is Zlan. Zlan means *God.* This name people have given me because, like God, I am able to create beautiful things with my hands." Some of Zlan's family members recalled how Zlan was named. Shortly before he was born, a relative dreamed that the child would have special abilities to create the human face or form (Dro 1986). Zlan is a name not uncommonly given to a Dan child believed to have been given supernatural qualities or abilities because of some strange happening during pregnancy or delivery (Kerser 1986).[2]

Zlan was born in Gangwebe, a We town in the Ivory Coast, on the River Cess which marks a portion of the boundary line between Liberia and the Ivory Coast. His father, Pei, was from Belewale (Kpeaple chiefdom), which is also a We town but on the Liberian side of the river. Pei was a carver who died when Zlan was very young. While still a small child, Zlan received a sign that he was destined to become a carver. An adze, a tool used for carving, fell out of an oil palm tree his uncle was cutting. Remembering the earlier dream, his mother set the adze aside, keeping it for him until he was ten or twelve. Practicing on *eddo,* a starchy tuber also known as taro, he carved faces. Later he practiced likenesses of faces on soapstone, softer and easier to carve than wood. Eventually the young Zlan began to carve wood.

In bush school at about the age of thirteen, Zlan carved his first mask. Afterwards in Belewale, where he lived for most of his life, he carved his first feast ladle. This was for a singer named Quiu who had commissioned a ladle with a handle that ended in the carving of a boy's head. With it the singer intended to take up an offering when he

[1]"Zlan" is spelled "Sra" in Himmelheber. Both approximate Dan-We pronunciation.
[2]When the child who is thought to have supernatural abilities is a girl, she is now often given the name Mary after the Christian mother of God, known through missionary influence. The Dan have no female deity.

performed in the villages (Himmelheber, *Negerkunst*, 1960, 173). Following this effort, Zlan carved many more ladles, and in this way learned to carve well (cat. nos. 14, 15, and 16).

As his fame began to spread, the chiefs from Dan, Mano, and We towns sent for him to commission carvings. Always he kept the special adze with him as a form of spiritual power and protection against witchcraft. Before he died he gave it to his younger brother, Kwege, who as a child had assisted Zlan and who then began to carve. Before Kwege died he handed down the adze to his son (or possibly nephew), Gblor, a carver of drums, who still keeps it in his possession (Kwege et al. 1986, in Belewale).

Zlan created new mask forms. He had a special "medicine" that enabled him to dream in an important way; because of this his dead father appeared to him in a dream and showed him at least one new mask form, the *blokila* with horns that jut forward as on cows, and, he said, two faces instead of one (Himmelheber, *Negerkunst*, 1960, 174), although he may have meant two sets of eyes.

He also carved full figures. For the Dan chief Bu in the Ivory Coast, he carved a female figure to look like his head wife. He carved her with a child on her back (as in cat. nos. 10, 11). For a wealthy man named Ba in the Ivory Coast, he carved a woman and a man (as in cat. nos. 12, 13), as well as ten rice bowls and a stool. He carved feast ladles and ornate gameboards for Chief Djudi of Toulepleu, Ivory Coast (fig. 22). He carved a gameboard with the head of a sheep for Diong Glalu of Bolu in the Ivory Coast, and for him he also carved ladles and masks.

Another person for whom he carved was the Mano chief Lama in Saniquellie, Liberia. He also carved a great drum for Qua, a rich man who also was father of the chief Utompie, other objects for Paramount Chief Mongru in Kanple, and especially many objects for Chief Tappe of Tapita (Himmelheber, *Negerkunst*, 1960, 174). In Toway Town he was frequently summoned by Chief Toway for whom he carved big rice bowls, a walking stick, feast ladles, small spoons, and gameboards. Toway's son, Old Man Blazua, remembered Zlan well and recalled that he was even given a wife who stayed in Toway Town to care for Zlan when he was there (1986). Zlan reportedly was summoned by many other chiefs of Dan, Mano, and We towns.

Payment for work was not agreed upon in advance. When the work was completed, Zlan was given gifts, which varied from a "garment and two boxes of rock salt," to "a cow and a second cow slaughtered in my honor" (Himmelheber, *Negerkunst*, 1960, 174-175).

Zlan had several wives, but one in particular is well remembered by the people of Belewale because she carved along with Zlan (interviews 1986, in Belewale). This is a very unusual situation in West Africa, where carving wood is a skill reserved for men and usually hidden from women. Sonzlanwon, whose name means "Snail, if God agrees," accompanied Zlan when he was called upon to work for a chief. The usual practice was for them to build themselves a secluded house in the forest near the town, work secretly there by day, and come to town at night. An apprentice who accompanied them cut the wood and Sonzlanwon began the carving work. She worked until

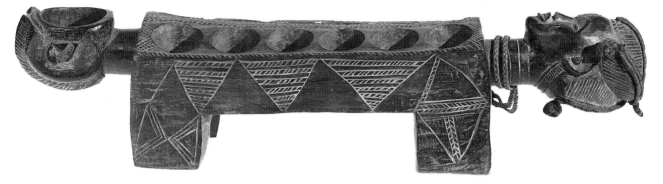

Fig. 22   Gameboard, probably by Zlan
See Himmelheber, *Negerkunst*, 1960, 45
Fischer and Himmelheber 1984, 140

the forms were ready in the raw, and then Zlan finished them (Himmelheber, *Negerkunst*, 1960, 175). According to several old people in Belewale who remembered her, she often finished a carving as well. She also carefully plaited and arranged the fiber tresses on Zlan's figures and feast ladles. In fact, when Zlan grew old and ill, she often carved under his direction. One old man stated that she carved more than Zlan himself (Tendee 1986). Her story is intriguingly incomplete, and unfortunately she died "a long time ago," but it is of interest that she has a successor in Zlan's niece, Younokwen, who is discussed below.

Zlan had many pupils who came to him in Belewale to learn carving through apprenticeship, and Belewale became known as a center for wood carving (Donner 1940, 81). Pupils frequently were sent by the chiefs of their villages in order that their village might then have a resident carver. Most pupils went to Zlan for a few days at a time to watch the master at work and carve under his supervision. Copying the master was the accepted practice, and, until a pupil could successfully imitate the master, he could not begin to develop his own style (Tompieme 1983). Finished pieces were brought to Zlan to be criticized. If he thought them inadequate, he threw them out. Sometimes Zlan finished a carving begun by a pupil, then sold it for his own profit. Considering this process, it is not surprising that Zlan's individual hand cannot always be recognized.

During the period of apprenticeship, which varied in length but often lasted many years, students worked on Zlan's farms and cut wood for him. Even after leaving the apprenticeship, for the rest of their lives former students traditionally brought gifts to thank and honor their master.

Several members of Zlan's family continue the carving tradition. Wrudugweh and Blekwa, nephews of Zlan, both believe they inherited Zlan's ability, although neither now spends much time carving. Blekwa has almost given it up as his eyesight has

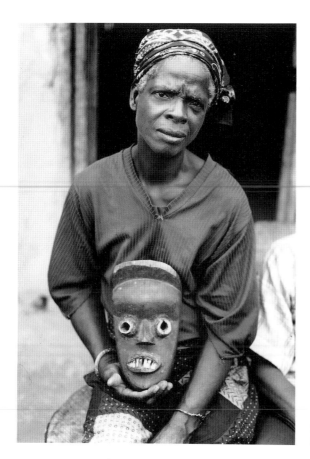

Fig. 24   Cheyorno Che Kwege, niece of Zlan
Belewale, February 1986
When she was a young girl her face served as a model for Zlan's sculpture.

Fig. 23   Younokwen in Yuopie Old Town with mask she carved
February 1986

become too poor. Younokwen (fig. 23), daughter of Zlan's sister, Ziate, claims to be
a carver as well. As a child she watched and learned from Zlan, and began to carve.
However, one night in 1961 (the year specifically remembered) the dead Zlan appeared
to her in a dream and warned her that she must stop carving wood or become barren.
As she was married and had several children at that time, she did put aside her carving
tools and went on to have a total of twelve children. Now, probably in her fifties, she
says she has started to carve again, although secretly because of widespread disapproval.
One of her sons, Pei, aged thirty-four, is an active carver who travels around to different
towns seeking commissions. His young son, Alfred, is also learning to carve (interviews
1986, in Belewale and Yuopie Old Town).

Figs. 25 and 26   Two figures collected in Belewale by 1937
Probably the work of one of Zlan's pupils
Rijksmuseum voor Volkenkunde, Leiden
2326–7 and 2326–8

Fig. 27 Zon carving, 1960
See Fischer, "Kunstler," 1963, 258
Fischer and Himmelheber 1984, 181

Zon, Master of the Herringbone Hair

Sometimes when you see a beautiful face, you remember that face and later carve it.
—Zon (1983)

Zon[1] was Zlan's pupil. Born into a Dan family of hereditary blacksmiths, Zon as a child learned that craft. For many years Zon was the only blacksmith in his town of Nuopie (Doe clan), where he forged tools and knives for the people. Although Zon says that he was "born with carving" (Fischer, "Künstler," 1963, 213), that is, he had innate ability, he never had carved wood other than tool handles until he was past forty.

In about 1948 or 1949, the We carver Zlan came to Nuopie and changed Zon's life. Zlan had been sent for in order to carve a mask for the village of Nuopie, but his lingering illness prevented him from carving when he arrived. Zlan first went to Zon's smithy and greeted him with the distinctive Liberian finger-snapping handshake, and addressed Zon as "father." When Zon asked him what he meant, Zlan explained that all the tools used in carving are made by the blacksmith; without the blacksmith Zlan could not carve. Zlan therefore considered Zon and all blacksmiths to be "father" to him (Fischer, "Künstler," 1963, 211). Zlan then asked Zon to make some carving tools as he had not brought any with him. When these were finished, Zlan and Zon went into the forest together to select the proper wood for the mask. After Zon had cut down a tree, Zlan told Zon, "Work it," and directed him. Thus, although Zon had no previous mask carving experience, under Zlan's tutelage he carved the required mask (Fischer, "Künstler," 1963, 212).

The mask Zon carved was a blua gle, a judge mask once associated with warriors. When Fischer, several years later, saw the same mask in a dance performance in Nuopie, it had risen in importance to become a gle va, big or great gle, the most important mask of a region. The mask itself has a look of authority. It has both human and animal features. The projecting tube eyes (nya gbo) are covered with metal discs; triangular cheekbones and a prognathous moveable jaw combine with wooden cow horns to give the mask a frightening nonhuman look. Eyebrows and a vertical tattoo line are depicted; decorative incising has been added to the hairline and to the jaw (fig. 28; note the similarity in another mask carved by Zon, cat. no. 18).

After this mask was finished, Zon went with Zlan to Tapita to carve for the Dan Paramount Chief Waipapee. They made "a lot of money," all of which Zlan kept, as well as the tools made by Zon. Zon returned to Nuopie and, as the only carver of his region, found his newfound skill in demand. He never again left Nuopie. When a chief or anyone else wanted to commission a carving, he or she came to Zon. The price was agreed upon in advance. Zon carved for the twenty villages of his Doe clan and for other towns as well.

He carved masks of various types and guardian heads (cat. no. 22), both of which are used for ritual purposes, and feast ladles for the traditional wunkirle (cat. no. 23), as well as similar pieces for nonritual, nontraditional purposes. According to Zon, feast

[1]Zon is spelled *Son* in Fischer. His name is the Dan equivalent of John.

ladles were sometimes wanted simply to decorate the home. Occasionally the patron ordered the piece to resell it, although the carver was never supposed to know this.

Zon also carved lime, figures of both men and women, which he assumed were to be displayed as works of art in the home (cat. nos. 24, 25). He said he was introduced to figure carving by his teacher, Zlan, who showed him how to make this form. Zon also carved gameboards, for which he usually earned four dollars, though he sometimes gave a "special price" (Zon 1983; cat. no. 26). He spoke of carving a gameboard for the great Chief Siawe, who was carried by hammock to Nuopie in order to commission the work. For that gameboard Zon received several goats in addition to money (Fischer, "Künstler," 1963, 213). Zon, on request, also carved various kinds of animal figures. Carving was satisfying work to him, but at times he had to lay it aside to forge the agricultural tools needed by the people of his village to clear the forest, farm, and harvest the rice.

Zon's style is distinctive and did not change greatly from his earliest works. The stylized human face that appears on most of his sculpture varies little. Although he seemed to believe that throughout his life he had carved individual, remembered faces, he had in reality created one face, copied repeatedly. Zon's treatment of the deangle face, and his sculpture in general, tends to be angular rather than rounded in form. Hair styles are done the same way, with one style for men, one for women. Typically, his carved pieces are carefully incised with a herringbone pattern. This might decorate the center of a gameboard, for example, representing the vertical tattoo markings of the forehead, or it is often used to embellish a carved coiffure.[2]

Zon traveled rarely, and maintained that he was the only carver in his region. He said that the only other carver he ever met was his teacher, Zlan, whose influence was brief although great. Other carvings may have influenced his work; nevertheless, he seems to have evolved an individual style that he had not been stimulated to develop during his carving career.

Dreaming apparently was not a source of inspiration and power for Zon, as it had been for Zlan, Dro, and Ldamie. Although Zon told Fischer ("Künstler," 1963, 213) that he rarely dreamed, he recounted one significant dream:

Occasionally, somebody comes to me but I do not see this person clearly. [Fischer injected a question about the gender of the person.] *I don't know whether it is a man or woman. The person wants to show me something new for my carvings. But I do not listen to this person. I act as I decide. No, I never encountered this person during the daytime. If I did, I would know who comes at night.*

Zon lived to be an old man, probably in his eighties, and died on 22 November 1985. He had no children and found no young man interested in learning his techniques of carving. In 1983 Zon gave us his last work as a farewell gesture upon our leaving Nuopie. It is a ma go, miniature mask, that had not been smoothed with leaves and stained, the usual finishing processes, and it bears the distinctive Zon face (fig. 29; cat. no. 21).

2"Master of the herringbone hair" is the author's term and was not used by the Dan people.

Fig. 28   Zon's first mask in performance, Nuopie Old Town, 1960
See Fischer and Himmelheber 1984, 89
Mask is now in the Museum für Völkerkunde, Hamburg

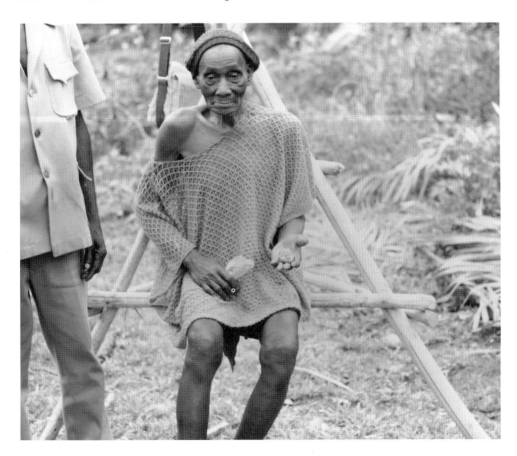

Fig. 29   Zon holding his last carving, a miniature mask, in 1983

Fig. 30   Dro
Tapita, February 1986

Dro, Carver of Today

I carve so that my name is spread in the country and the people ask, "Who has made this?"
—Dro (Himmelheber, *Negerkunst*, 1960, 188)

Dro, who is approximately sixty-four years old, is the grandnephew of the carver Zlan. Dro's father, Lopa, also was a carver but was known only as a carver of rice bowls.

Until recently, Dro lived and carved in the town of Kpeaple (Kpeaple chiefdom), on the River Cess. Although he is We, as he comes from the Dan-We border area he typically speaks fluent Dan as well. He also speaks fluent English, having attended an American missionary school for eight years.

Dro knew from the time he was very young that he was a carver. He explained, "I was born with it. It is in my head" (Dro 1986). When he was seven years old he carved his first mask:

It was such a small one. It came out well at once. When I had finished it I showed it to my father. He said, "Who taught you this?" I answered, "I learned it from you" (Dro, in Himmelheber, *Negerkunst*, 1960, 179).

After that he practiced carving on tree fungus, which was soft and readily available. He laughingly recalls how his mother frequently became angry with him when he brought his messy carving work into the house; however, he was defended by his father, who had dreamed that Dro was to become a carver, and was allowed to continue. Dro learned to become a "professional" carver at the time when he worked at the Firestone rubber plantation, and was called to learn the skill by Zlan, who was getting old (Dro 1983 and 1986). Zlan later named Dro one of his five best pupils (Himmelheber, *Negerkunst*, 1960, 177).

Dro felt that he first gained recognition when Himmelheber came to his village in 1957 and saw his work. At this time Dro was carving in a secluded place in the forest, marked by a curtain of bast, which prohibits entry. Thus he carved sacred objects in the traditional manner. In 1957 he went to Tapita (Gblar clan) to a Baptist missionary school for eight years. He supported himself by carving for foreigners from the Firestone rubber plantation and his name became known over a wide area. Himmelheber deplored the coarse masks fitted with embellishments of hide that Dro carved at this time for the "young gentlemen" of Firestone. However, Dro explained that this is how they wanted them (Himmelheber, *Negerkunst*, 1960, 175). He also carved for missionary people, other foreigners, as well as for traditional purposes in the Dan villages around Tapita.

In 1965 Dro returned to his family in Kpeaple. He and his wife, Gaye, had ten children, six of whom have survived to adulthood. To earn a living his family grew coffee and cocoa as cash crops, and rice and other foods for their own consumption. He earned a little extra cash by carving, sitting under a tree by the side of the road for all to see, and selling his carvings. He carved traditional art forms, including feast ladles carved on commission for various klawaiyno (the We equivalent of wunkirle). Dro recalled three particular ladles over the past five years from towns in the vicinity of

Kpeaple, one of them from the town of Klau in the Ivory Coast, where he and his wife were honored with a feast for which a goat was killed. All of these were paid for in rice and an additional small amount of money (Dro 1986).

Dro became despondent when his wife died in 1985. People who knew him said all he did was grieve. Finally he decided to make the move back to Tapita where he had Baptist missionary friends. He left the farms and house in the hands of his children in Kpeaple. Recently he took a young wife, and has become a church-going Baptist. The market for his carving has become even more alien to his traditional culture, centering on the missionaries and other foreigners in the area.

Dro believes in the importance of dreams. He expressed the belief that "dreams tell you what God has already made you." Before he went to Christian school he remembered seeing in dreams, and later carving, new and different masks of types he had never seen before. He also recounted one recurring dream in which he ran from a pursuing mask-spirit of a fearsome bugle type with projecting tube eyes: "The devil [the common Liberian English term for gle] wants to come to me so he can be a dancer. I never wanted to take it." Although he resisted wearing the mask that appeared to him in the dream in village ritual, he has since carved that mask many times (Dro 1986).

Dro carves masks (cat. no. 27, 28), figures, feast ladles (one of which he gave to the National Museum in Monrovia), gameboards (cat. no. 30), and also newer forms such as animals. He is experimenting with elephant tables and eagles and proudly claims that he can carve anything.

None of Dro's children carve; in fact, he says it is hard to find any young person interested in carving any more. They tell him there is no money in it and laugh at him, asking him who will buy the things he carves. Nevertheless, he has several assistants, teenaged boys who, when not in school, run errands for him, do some of the sanding of the carved objects, and sit with him for hours, conversing and watching him carve.

Dro Carves a Feast Ladle

On Friday, 31 January 1986, in order to compare Dro's present-day methods of carving with the traditional ones, I commissioned him to carve a traditional object, a feast ladle with a woman's head on the handle (figs. 31–42; cat. no. 29).[1] A price was agreed upon, which is his usual procedure, unlike the former practice of the client determining the worth upon receipt of the finished carving.

First, he went into the forest and with a machete cut down a *yuon,* a wild rubber tree. It is one of the two varieties of wild rubber that he regularly uses, both having medium-soft, not very dense wood. Had he been working in the traditional way, he said it would have been necessary to offer a sacrifice to the tree, such as a white chicken, before cutting it down. After lunch he removed some of the bark from the wood and marked the correct proportions with charcoal. Working with an axe, he roughed out the general shape. Formerly he would have used the traditional adze for this. Again using charcoal, he marked the overall shape. With a chisel and mallet he hollowed out the bowl of the ladle and the narrow neck.

The next day, Saturday, 1 February, Dro spent the whole day carving the finished shape of the ladle, including the features of the face, the incised coiffure, the neck with neck rings, and the bowl that comes to a point at the bottom, to which he gave a vertically faceted back with incised designs. He did all this with a chisel and a small knife. He used a broken piece of glass to smooth the interior of the bowl and a metal file to smooth the whole piece. He then handed it to one of his young assistants who spent about two hours smoothing it with sandpaper. Formerly a special rough leaf from a local shrub was used to sand a carved piece. Dro then darkened the piece by applying hair dye he had bought in town. The old way to darken a piece, he said, was to bury it in the mud of the swamp for many days. Other carvers spoke of a concoction made from a certain boiled but unnamed leaf that turned the wood dark on application.

Dro does not work on Sundays, reserving that day for rest and church. On Monday, 3 February, he finished the ladle by first polishing and shining it with commercial shoe polish. He then cut up a white plastic bottle into small rectangles that he inserted for teeth after cutting small incisions into the mouth in several places. Formerly palm oil might be applied, and animal teeth or metal pieces would be used for teeth. Finally he painted white kaolin across the eyes in the cosmetic fashion of the women on feast days. This was a traditional step.

[1] For a fuller understanding of the steps in carving by older, more traditional Dan carvers, see Fischer, "Künstler," 1963 and Fischer and Himmelheber 1984, 171-181.

Fig. 31   Dro fells a wild rubber tree in the forest, . . .

Fig. 32   . . . carries the log back to his workshop, . . .

Fig. 33   . . . and begins to shape the log with an axe

Fig. 34   He marks one side of the log with charcoal
to indicate the proportions of the feast ladle

Fig. 35   With the axe Dro carves the rough shape of the feast ladle

Fig. 36   He continues to carve the shape of the feast ladle …

Fig. 37   … after marking the forms with charcoal

Fig. 38   Dro uses a mallet and chisel to refine the shape …

Fig. 39   ...and carves the facial features with a knife, ...

Fig. 40   ...stopping frequently to look at his work

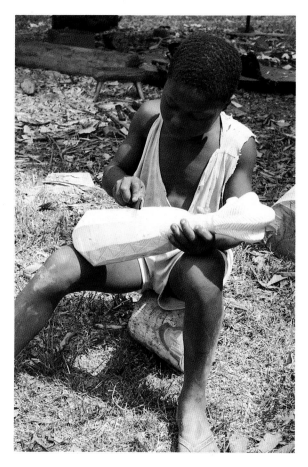

Fig. 41   Dro's young apprentice smooths the feast ladle with sandpaper

Fig. 42   Dro blackens the feast ladle with hair dye

Fig. 43   Ldamie
Donner 1940, 47

Ldamie, Figurative Brass-Caster

Ldamie had a skill gotten by dreams. He had a God-given gift. Nobody taught him.
—Yanemie Ldamie, son of Ldamie (1983)

Ldamie[1] was born in Zonleh, in northeastern Liberia. He was from a long line of brass-casters who dated back in memory to his great-grandfathers. As a boy Ldamie spent hours watching an older male relative, Munmie, who was a master brass-caster. From Munmie, Ldamie learned the skills to become a blacksmith, the usual first step to becoming a brass-caster. He tried his hand at wood carving, too, and eventually brass casting.

Although Ldamie may have learned the techniques of brass casting from Munmie, his artistic ability and real knowledge was obtained through dreams, according to Dan belief and to the people who remembered Ldamie. According to traditional Dan belief, dreaming is the mechanism through which dü, the spiritual power essential to becoming great in any area, communicates with individuals.

As a child Ldamie migrated with his family to the nearby town of Gaple (Gbehlay clan), where he lived the rest of his life. He married and, with his wives, cultivated rice and other foods. Unable to earn a living from their art alone, artists of this area have almost always had to be farmers in order to survive.

We know that Ldamie cast brass swords, sword handles and sheaths, gun handles, fly whisk handles, jewelry, bells, and animals. He also became one of the few figurative brass-casters of the region. He began producing brass figures for an occasional chief or wealthy man, figures that were considered status symbols, to be brought out and placed by the fire where they would catch the eye of a distinguished visitor. Ldamie observed village life around him and portrayed in genre figures many daily activities.

According to Ldamie's son, Ldamie's big break came one day when a government official visited Gaple with a broken typewriter. Although Ldamie could not read or write, he proved his metallurgical skill to the official by repairing the broken typewriter. After that he began casting brass figures for officials to take home with them to the national capital, Monrovia. His name spread and he became famous.

Eventually the government had a workshop built for Ldamie, and he became "so rich and famous that he was carried everywhere in a hammock [a sign of prestige], and never again had to walk" (Yanemie Ldamie 1983).

The dates that Ldamie lived and cast are not definitely known. However, all sixteen of the pieces in Ldamie's style for which we have some collection data were collected in

---

[1]Etta Becker Donner, a German anthropologist who spent some time in northeast Liberia during the 1930s, wrote about brass casting in the Dan-We region (Donner 1940, 47). She mentioned a Dan brass-caster from Gaple and included a photograph of him, but never named him. She also mentioned that he was one of only two brass-casters in northeast Liberia who made figures from brass.

In March 1983, nearly fifty years after that photo was taken, we took the photograph to Gaple and had it identified as Ldamie by his son, Yanemie, and a group of old people. It also was verified by Gbangor, an oral historian in the neighboring town of Beple. We asked many questions, showed photographs of brass pieces, and from both sources obtained some information about Ldamie and his works although he had died "a long time ago," probably in the 1950s. Again in February 1986 Ldamie was remembered and discussed by Chief Blazua, son of Chief Toway in Toway Town, for whom Ldamie had cast many brass objects.

Liberia in the 1920s and 1930s. An exception is cat. no. 41, which was collected in 1986. Others for which data is lacking may have been collected later than the 1930s. Donner referred to Ldamie as an "old man" in the 1930s, possibly referring to his social status rather than his age. Yet his son, Yanemie, who gave his age as twenty-nine in 1983, was born in 1953-1954. He said his father died when he was a small child, therefore probably in the 1950s. We can assume that Ldamie cast figures in the 1920s and 1930s; he may have begun before that time but the latest possible dates that he might have worked would be the 1950s.

Donner described Ldamie as a "genuine craftsman who obstinately sticks to old complicated subjects and carefully hammers the metal smooth after it has cooled" (Donner 1939, 149). Significant characteristics of Ldamie's figures, with some variations, include the shape and proportions of the features of the face—the eyes, nose, mouth, and ears, the shape of the body, particularly the legs and back. Legs are portrayed with muscular calves; backs usually have a vertical depression to indicate the spinal concavity. Hands and feet are usually flat and rigid with fingers and toes indicated by incised lines, and include the details of finger- and toenails. Sprues generally have been cut off from the bottoms of the feet. All the figures appear to have been hammered after casting, producing an attractive faceted surface that catches the light. Ldamie's attention to detail was remarkable for such features as hairstyle, the woven pattern of a winnowing basket, details of jewelry, or decorations on a stool; all detail was added with strands of wax that had been braided, twisted, or spiraled. The height of each figure varies from about 8 to 10 inches, although there is one group figure of trumpet players in the Peabody Museum that measures only about 6 inches (13.5 centimeters) in height. It is, however, about 8 inches (18 centimeters) wide, and undoubtedly required as much brass to make as some of the single figures (Schwab 1947, fig. 68s).

Often repeating the same themes, Ldamie depicted people in action or carrying objects of importance. A number of figures exist of standing women with outstretched hands, sometimes holding ceremonial knives, staffs, or other ritual objects (cat. nos. 41, 42). These women are elaborately coiffed and display bracelets, necklaces, waistbands, and/or knee rings, although they are otherwise unclothed. Women also are shown performing such daily chores as carrying water, pounding or winnowing rice (cat. nos. 38, 40), or tending babies. Men are often shown as blacksmiths (cat. no. 39) or musicians. One figure is of a carrier with a load on his head, another is of a prisoner, his hands tied behind his back. Islamic influences are also seen. For example, one figure depicts a chief carrying a walking stick and wearing an Islamic amulet hat. Showing western influences, another figure portrays a man carrying a musket who wears a European-style top hat; another is of a man wearing an Islamic amulet hat, seated in a folding chair, and about to strike the match in his hand to light the pipe in his mouth (cat. no. 44).

Continuity and Change

All four of the sculptors whose work is represented in this exhibition were innovative and creative artists. Three of them developed the ability to work with different materials—Zlan, Ldamie, and Zon working as blacksmiths and sculptors. Ldamie, as well, was a brass-caster. Another common factor for them as artists was dreaming as an important source of inspiration, except for Zon who refused to accept new forms in his dreams. All three of the carvers worked when inspired, even without a commission, which suggests their artistic natures and love of carving. But for each of them, perhaps, the gain of prestige and honor was the overriding motive to pursue their art.

Ldamie and Zlan, both of whom sculpted during the first half of the twentieth century, were honored with gifts that indicated their prestige. This was the accepted means of payment for carving or casting, rather than an agreed-upon cash price. Having completed some carving for a particular chief, Zlan told of being presented with a cow and having a second cow slaughtered in his honor (Himmelheber, *Negerkunst* 1960, 174). This meant a large feast was arranged by the chief, attended by all the people and the most important masks of an area, with Zlan as guest of honor. Ldamie was honored by the government officials who arranged for troops to build him a workshop, and by the prestige of being publicly carried from town to town in a hammock. Zon, who began carving in 1948 or 1949, also carved for chiefs and was no doubt honored appropriately. However, there were more foreigners in Liberia by this time, bringing changes in traditions and the "introduction of western civilized ways."

The power and prestige of the chiefs were beginning to diminish, and new status symbols from the West began to supplant the traditional Dan sculpted ones. Universally, there was a greater dependence on money, which could buy imported goods from the outside. Zon mentioned carving a gameboard for a chief who paid him in money, but who also honored him in the traditional way with a gift of several goats (Fischer, "Künstler," 1963, 213). Dro today seeks recognition through his carving from the Dan people in the area, and from the missionaries, traders, and other foreigners in Tapita. He sells his carvings or accepts commissions with the price agreed upon in advance, and in this way earns money, but no great prestige.

We note differences in the quality of the sculptures of these three carvers of three generations. Zlan's work is among the finest the Dan have ever produced. It was forcefully yet meticulously conceived and carved, with attention to detail. Zon's carving is well executed, although it tends to be less fluid and somewhat repetitive. Dro's work seems to be carved with less confidence and care, and his attention to detail is not as great.

It seems possible that the main reason for the decline in quality relates to the changes in the system of patronage and prestige. Zlan carved what he considered to be "important" pieces for discerning, powerful chiefs or individuals in his culture who knew the traditional forms and their meanings, insisting upon the best. Zon, to some extent, and Dro, largely, carved for a new, external market. Many pieces were bought by traders to be resold to outsiders, or were sold directly to missionaries, Peace Corps volunteers, rubber plantation workers, or other foreigners. These people, who did not know the artistic tradition, were less demanding; they merely wanted souvenirs, not objects important to their station in life, as was true for the chiefs. Less work was required to produce a saleable piece for good money.

Even their dreams, and the effect of those dreams on the sculptors, reflect the changing culture. The Dan culture has been a conservative one, controlled by the elders and placing value on continuing the traditions of the past. Dreaming was the means of visualizing the traditional art forms. Even the innovations seen in dreams were well within the mainstream of the culture. Zlan and Ldamie were greatly influenced by dreams, both in envisioning their creations and receiving the necessary spiritual power, dü, to create. Zon, who carved later, at a time of greater change, resisted the message of his dreams by refusing to see what the stranger wanted to show him, and Dro, who had abandoned his culture, ran from the traditions he saw represented in his dreams.

Perhaps, too, because carving today no longer involves the prestige it once did in the villages of the Dan, many of the creative young men have gone elsewhere, particularly to the cities, to learn new skills and try something different to earn money. There are fewer of the old Dan carvers left today to provide the continuity once required by the whole Dan culture. Inevitably, as the culture changes with the times, the art will change too.

Catalogue of the Exhibition

The Tradition of Masks

Probably the earliest sculptural forms of the Dan, masks are also the most numerous and culturally the most important. From the established tradition of masks came other sculptural forms that frequently incorporated the face of the deangle mask and developed from that tradition.

1. *Deangle* mask
Wood, 22 cm
Janine and Michael Heymann, San Francisco

The deangle face was the inspiration for much Dan art.

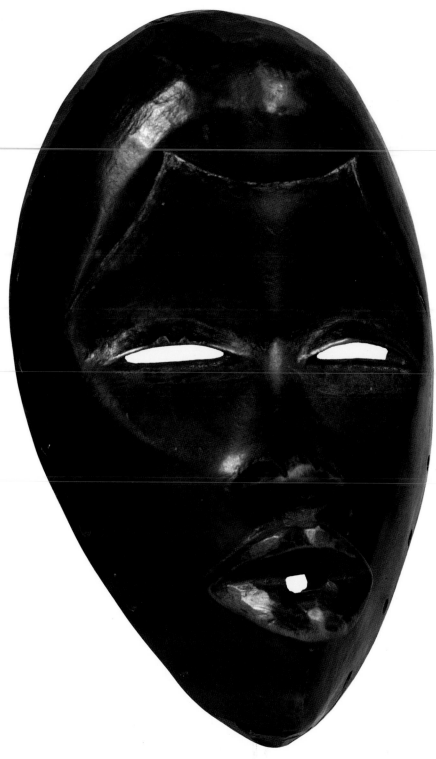

2. *Deangle* mask
Wood, 19 cm
Ms. Penelope Evans, San Francisco

The deangle face represents an idealized
form of feminine beauty and grace to the
Dan.

3. *Deangle* mask
Wood, vegetable fiber coiffure, 24.2 cm
Peabody Museum of Archaeology and Ethnology,
Harvard University. 37–77–50/2672

This mask spirit was known for its fine
dancing and singing. It was said that when
it was entertaining, townspeople put aside
their work and travelers went no further
but sat down to watch and listen (Harley
1941, 19. Also see Harley 1950, pl. IVa;
Adams, "African Treasures," 1982, 31). Collected by Harley in 1936 from Toway
Town.

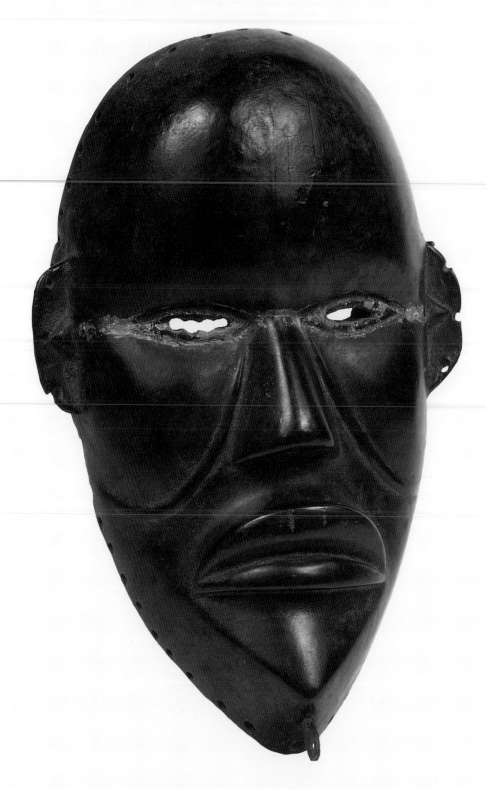

4. Mano *deangle* mask
Wood, 26 cm
William C. Siegmann collection

The essential functions of the deangle are
nurturing, teaching, and/or entertaining.

5. *Bugle* mask
Wood, 22.1 cm
Peabody Museum of Archaeology and Ethnology,
Harvard University. 37-77-50/3018

At rice-planting time, all hoes were brought
into the presence of this mask. The mask
was not worn at this time, but prayed to
and washed in a pan of water in which a
puff adder, a nonvenomous snake, had been
added. Then all the men drank the water
and abstained from food, drinking only
palm wine until the planting was finished
(Harley's unpublished notes, 1937, Peabody
Museum). Collected by George Harley in
1936 or 1937.

6. *Kagle* mask
Wood, 21 cm
Marc and Ruth Franklin, San Francisco

Kagle rouses the people with its erratic,
aggressive behavior. Its name derives from
the hooked sticks (ka) it carries and flings
into the crowd. Kagle may take a number
of animal forms; this form, the chimpanzee,
is most often represented (see Schadler
1975, 90; and 1976, 31).

7. *Gle va* mask
Wood, porcelain, iron, rope, 33 cm
New Orleans Museum of Art
Bequest of Victor K. Kiam. 77.248

A gle va is the most important mask spirit of
an area. Its main function is to settle important disputes and make peace between villages. To emphasize its importance, it is
often larger than the other masks, and physically suggests its power.
In this case its facial planes protrude and
its animal jaw moves, allowing speech,
(see Fagaly 1983, 29).

Zlan, Great Master and Teacher

8. *Deangle* mask, probably by Zlan
Wood, paint, metal inserts for teeth, 24.8 cm
Peabody Museum of Archaeology and Ethnology,
Harvard University. 48–36–50/7377

This mask was collected by George Harley
in Belewale in 1947 or 1948. Its name was
Di kela, and it was said to dance for vic-
torious warriors (Harley 1950, 26).

9. *Gle va* hippopotamus mask,
probably by Zlan
Wood, 32 cm
Private collection, Switzerland

This humanized hippopotamus mask was
meant to look powerful, combining human
features with those of a strong and powerful
animal. Its function was that of a gle va, or
great mask spirit, which served as a judge
mask. When it appeared in the village it
spoke through its moveable jaw, constantly
clattering its upper and lower jaw together.
This mask was collected by Himmelheber
between 1949 and 1953, (see Himmelheber,
Negerkunst, 1960, 141; Leuzinger 1972, 106;
Fischer and Himmelheber 1984, 97).

Fig. 44   Fanta Togbeh and her baby
Nyor Diaple, February 1986

10. Mother and child figure,
probably by Zlan
Wood, vegetable fiber coiffure,
metal inserts for teeth, 63 cm
Musée National des Arts Africains et Océaniens, Paris.
MNAN 1963–163
San Francisco only

The number 22, inscribed in green at the
back of the right leg, confirms that this figure was exhibited in the army pavilion at
the Colonial Exposition in France in 1931.
All of the objects from the exposition were
put in this museum, formerly known as the
Musée des Colonies. This piece was collected in the Touba region of the Ivory
Coast where there was a French military
post. See also Meauzé 1968, 55.

11. Carved figure with baby on back
(front and back views),
probably by Zlan
Wood, vegetable fiber coiffure, metal inserts for teeth, 64 cm
Jacques Kerchache collection, Paris

Zlan was commissioned by Dan Chief Bu
to carve a portrait of his head wife with her
baby on her back. Both this piece and cat.
no. 10 answer to this description. For the
carving, Zlan was honored with gifts of a
sheep, a goat, and a large garment (Zlan, in
Himmelheber, *Negerkunst*, 1960, 174).

12 and 13. A pair of figures, probably by Zlan
Wood, vegetable fiber coiffures, metal inserts
for teeth, cotton loin cloths, 58 and 60 cm
Private collection

Note particularly the elaborate vegetable
fiber coiffures on this male and female pair.
They were collected in northeastern Liberia
by Alfred Tulk, an artist friend of the missionary George Harley. Tulk, with his wife
Ethel, spent 1932 and part of 1933 in Liberia
living with the Harleys and painting
(see Schwab 1947, figs. 72e and f).

14. *Wunkirmian*, feast ladle, by Zlan
Wood, vegetable fiber coiffure, 70 cm
Private collection, Switzerland

Feast ladles such as this distinctive double-
bowled one were carried as a badge of
honor by the wunkirle, the most hospitable
woman of a village quarter. This one was
collected by Himmelheber between 1949
and 1953, (see Himmelheber, *Negerkunst*,
1960, 163; Fischer and Himmelheber
1984, 128).

15. *Wunkirmian*, feast ladle,
probably by Zlan
Wood, vegetable fiber coiffure,
aluminum inserts for teeth, 70 cm
Karob collection, Boston

This feast ladle was collected in 1933 by
Alfred Tulk in Blimiple, Liberia, near
Butuo, after bargaining with the woman
who owned it, and her husband (see Adams,
Designs, 1982, 97–98).

16. *Wunkirmian*, feast ladle,
probably by Zlan
Wood, vegetable fiber coiffure, metal inserts
for teeth, 68.6 cm
New Orleans Museum of Art,
Bequest of Victor Kiam. 72.277

This Janus-faced ladle portrays two different
faces in the same style, (see Fagaly
1983, 33).

Zon, Master of the Herringbone Hair

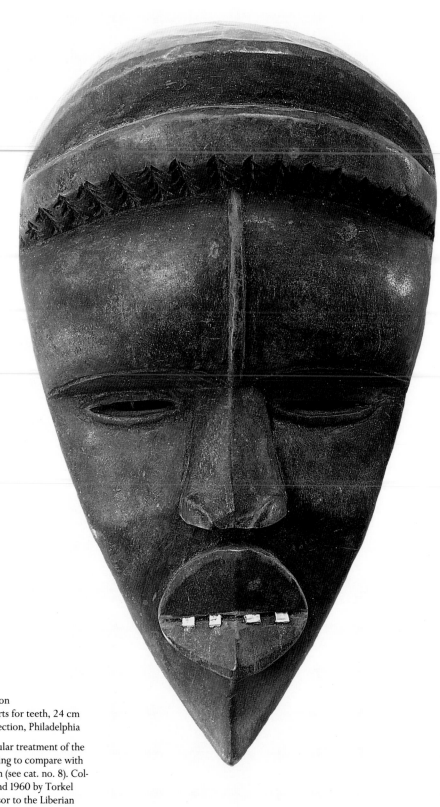

17. *Deangle* mask, by Zon
Wood, with metal inserts for teeth, 24 cm
Edward J. Biggane collection, Philadelphia

Zon's distinctively angular treatment of the deangle face is interesting to compare with that of his teacher, Zlan (see cat. no. 8). Collected between 1954 and 1960 by Torkel Holsoe, a forestry advisor to the Liberian government under the auspices of a U.S. government technical assistance program.

18. Judge mask, by Zon
Wood, paint, porcelain, cowrie shells,
fur, hide, vegetable fiber, 40 cm
Private collection, Washington, D.C.

This mask is similar to the first mask Zon
carved under Zlan's tutelage. Both combine
bush cow and human features (see fig. 28).
Collected in Diala, Liberia, by Torkel
Holsoe, along with cat. nos. 19 and 20
between 1958–1960. All three were said
to be judge masks.

19. Judge mask, by Zon
Wood, paint, porcelain, carpet tacks,
metal hooks, red beads, 47 cm
Private collection, Washington, D.C.

This mask represents the humanized features of a bird, possibly the hornbill, which
is the mythological bringer of the oil palm
among the northern Dan. Collected with
cat. nos. 18 and 20.

20. Judge mask, by Zon
Wood, paint, porcelain, fur, aluminum,
metal hooks, 51 cm
Private collection, Washington, D.C.

The third of the three collected in Diala,
Liberia, between 1958–1960, this mask
represents a gle in crocodile form, with
multiple sets of eyes.

21. *Ma go*, miniature mask, by Zon
Wood, 8 cm
Private collection

Collected in 1983 from Zon in Nuopie, this
unsmoothed and unstained miniature mask
was said to be his last carving.

22. Janus-faced guardian head, by Zon
Wood, metal inserts for teeth, 31 cm
William C. Siegmann collection

This head, carved in the round, incorpo-
rates two deangle faces and is on a flat-
bottomed pedestal base. Guardian heads
are intended to provide their owners with
spiritual protection. It was collected from
a Mandingo dealer in Liberia in 1984.

23. *Wunkirmian*, feast ladle, by Zon
Wood, metal inserts for teeth, 58 cm
Museum für Völkerkunde, Basel. 17698
San Francisco only

The hairstyle depicted on the head of this
ladle is carefully carved in a herringbone
pattern, characteristic of Zon's style. Collected from Zon by Eberhard Fischer in
Nuopie in 1960, it had been carved for a
specific wunkirle who was persuaded with
a gift to let Zon sell it instead to Fischer,
(see Fischer, "Künstler," 1963, 215; Fischer
and Himmelheber 1984, 186).

24 and 25. A pair of figures, by Zon
Wood, 55 cm
Edward J. Biggane collection, Philadelphia

Lime, wooden figures carved to portray real
people, were prized as status symbols by
their owners. Collected between 1954 and
1960 in northeastern Liberia by Torkel
Holsoe.

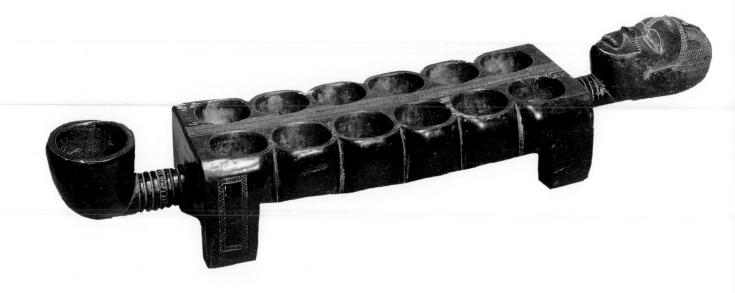

26. Gameboard, by Zon
Wood, 80 cm long
The Fine Arts Museums of San Francisco. 78.89

The game ma kpon, although ancient, is still
popular among the Dan. Great care is taken
in carving these boards so that they may be
handed down to future generations.

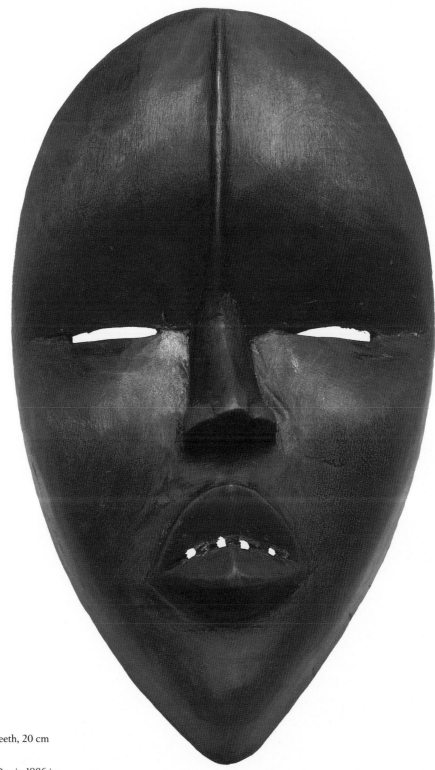

27. *Deangle* mask, by Dro
Wood, plastic inserts for teeth, 20 cm
Private collection

This mask was carved by Dro in 1986 in
Tapita, Liberia. It was made to sell, not
commissioned. After the price was agreed
upon, white plastic teeth cut from a bottle
were inserted and the mask was given a
shoe-polish shine.

28. *Kagle* mask, by Dro
Wood, paint, 26 cm
Edward J. Biggane collection, Philadelphia

This mask abstractly portrays a chimpanzee, frequently depicted on kagle masks. It was collected in northeastern Liberia by Torkel Holsoe between 1954 and 1960.

29. *Wunkirmian*, feast ladle, by Dro
Wood, plastic inserts for teeth, 50 cm
Private collection

Dro was commissioned to carve this feast
ladle on 31 January 1986. He worked on it
for three days. See photographs of the various carving stages in "Dro, Carver of
Today."

30. Gameboard, by Dro
Wood, metal inserts for teeth, 75 cm long
Private collection

The head on the end of a gameboard is
sometimes intended to be a portrait.

Brass Casting

The practice of casting brass in the forms of jewelry, ritual objects, and various kinds of miniatures was a well-established tradition from which figurative brass casting evolved.

31. Miniature mask
Brass, 12 cm
William C. Siegmann collection

The deangle face is seen in brass miniature masks, such as this one, which is cast with fine detail in the cire-perdue or lost wax method of casting.

32. *Nitien*, sacred ring
Brass, 11.1 cm in diameter
Peabody Museum of Archaeology and Ethnology,
Harvard University. H990

Niatien (plural form) are used as objects of
divination. They are consulted at times of
uncertainty, and are said to move about
during the night. The position in which they
are found in the morning indicates their
answer (Schwab 1947, 404–415; 363,
fig.67v). This nitien was collected by
George Schwab in northeastern Liberia in
1927.

33. Necklace
Glass and brass, restrung on palm fiber string
40 cm long
Private collection

The brass leopard teeth on this glass bead
and brass necklace are symbols of prestige.
It was collected in northeastern Liberia by
Alfred Tulk in 1932 or 1933.

34. Pair of knee rings
Brass, each 14.1 cm in diameter
Peabody Museum of Archaeology and Ethnology,
Harvard University. H215

Knee rings were worn by women just below
the knee, usually in pairs, (see Schwab 1947,
fig. 671). Collected in northeastern Liberia
by George Schwab in 1927.

35. Container
Brass, fiber string, height 14 cm, diameter 4 cm
Peabody Museum of Archaeology and Ethnology,
Harvard University. 29–76–50/H1004

This finely worked container, made by the
cire-perdue method of brass casting, is
thought to represent a palm fiber quiver for
arrows, made in miniature as a status
object. It was collected in northeastern Liberia by George Schwab in 1927.

36. Anklet
Brass, 14 cm x 11.5 cm diameter
Peabody Museum of Archaeology and Ethnology,
Harvard University. 37–77–50/2665

This anklet, cast in a single piece, is made
up of many tiers of little bells, which added
music to each step of its wearer (see Schwab
1947, fig. 66c). It was collected in northeastern Liberia by George Harley in 1936 or
1937.

37. Bell
Brass, 11 cm
Private collection, San Francisco

This bell was collected by Alfred Tulk from
the chief of Gampa (Ganta) in 1932. He was
warned by the chief that the bell had the
power to speak and could provoke dire
consequences. Tulk silenced the bell by
tying up its tongue.

Ldamie, Figurative Brass-Caster

Fig. 45   Girls pounding rice in mortar and pestle
Kanple, January 1986

38. Two young women with mortar and pestles,
probably by Ldamie
Brass, 17 cm
Lowie Museum of Anthropology,
University of California, Berkeley. 5–12978

Pounding rice in order to hull it is a common sight in Dan villages. These two young
women are pounding in turn (see Bascom
1973, 63).

39. Blacksmith, probably by Ldamie
Brass, 16 cm
Karob collection, Boston

The blacksmith is a respected personage among the Dan. It is he who has always forged the tools for farming, and formerly the weapons for defense. Brass-casters were blacksmiths as well. This figure was collected in 1933 by Alfred Tulk from George Dunbar, district commissioner, northeastern Liberia. Dunbar received this and a number of other brass figures from village chiefs in lieu of hut taxes.

Fig. 46   Girl winnowing rice
Kanple, January 1986

40. Woman with winnowing basket,
probably by Ldamie
Brass, 21 cm
National Museum of Natural History
The Smithsonian Institution, Washington, D.C. 398838

Ldamie's brass figures often show people
engaged in daily village activities, such as
this woman who is winnowing rice before
cooking it in order to get rid of the chaff
(see Sibley and Westermann 1928). Collected before 1928.

41. Woman with leopard tooth necklace,
probably by Ldamie
Brass, 20 cm
William C. Siegmann collection

This figure was collected in a small town
in northeast Liberia in 1986. Although it
probably was cast around the same time as
the other Ldamie pieces, the different
patina results from remaining in use in
a Dan town.

42 and 43. A pair of figures,
probably by Ldamie
Brass, 18.5 and 19 cm
Private collection

The woman holds a nitien, sacred brass
ring, in one hand and a clitoridectomy knife
in the other. The man holds a knife and a
medicine horn. Collected at the same time
as cat. no. 39.

44. Seated man with pipe and matches, probably by Ldamie
Brass, 17 cm
Private collection

Outside influences are evident in this figure. It depicts a man wearing an Islamic hat, seated in a folding camp chair, and lighting his pipe with a match. Also collected with cat. nos. 39, 42, and 43.

45. Elephant, possibly by Ldamie
Brass, 30.5 cm
Private collection

This fanciful depiction of an elephant illustrates that elephants were remembered, although few were left in northeastern Liberia by 1933 when this piece was collected. Also collected with cat. nos. 39, 42, 43, and 44.

Bibliography

Interviews Conducted by Barbara C. Johnson in Liberia, 1983:

Dro Lopa, carver. In Kpeaple, 14 March 1983.

Gaye Wanyen, brother of the late carver Yayua. In Ma Diaple, near Nyor Diaple, 22 March 1983.

Gbangor Gweh, oral historian. In Beple, near Gaple, 23 March 1983.

Kerser, Peter, Dan research assistant. In Monrovia, 26 March 1983.

Si Gbar, carver. In Barluople, near Nyor Diaple, 19 March 1983.

Tompieme Whien, carver. In Nyor Diaple, 20 March 1983.

Woto Mongru, paramount chief. In Kanple, 23 March 1983.

Yanemie Ldamie, son of the brass-caster Ldamie. In Gaple, 19 March 1983.

Zerlee Kenkal, carver. In Butuo, 16 March 1983.

Zon Tia, carver. In Nuopie, 13 March 1983.

In 1986:

Blazua Toway, chief and son of Paramount Chief Toway. In Toway Town, 2 February 1986.

Blekwa Kaska, carver, nephew of Zlan. In Belewale, 5 February 1986.

Dro Lopa, carver. In Tapita, 31 January-1 February 1986.

Kerser, Peter, Dan research assistant. 28 January-13 February 1986.

Kle Sayzee, carver. In Yuopie Old Town, 5 February 1986.

Kpor, Wonsieh Anthony, young man. In Yuopie Old Town, 4 February 1986.

Kwege, Cheyorno Che, niece of Zlan. In Belewale, 5 February 1986.

Pei Glaweh, carver, son of woman carver. In Yuopie Old Town, 5 February 1986.

Tabmen, George W. [George Wowoa Tame-Tabmen] In Monrovia, 12-13 February 1986.

Tendee, blacksmith and elder. In Belewale, 3 February 1986.

Wrudugweh, carver and nephew of Zlan. In Belewale, 4 February 1986.

Younokwen, woman carver, niece of Zlan. In Yuopie Old Town, 4-5 February 1986.

Adams, Monni. "African Treasures of the Peabody Museum, Harvard," *African Arts* 15, no. 4 (August 1982):31

_____. *Designs for Living,* exh. cat. Cambridge: Carpenter Center for the Visual Arts, Harvard University, 1982.

Bascom, William. *African Art in Cultural Perspective.* New York: W.W. Norton and Co., Inc., 1973.

Biebuyck, Daniel P., ed. *Tradition and Creativity in Tribal Art.* Berkeley: University of California Press, 1969.

Brincard, Marie-Thérèse, ed. *The Art of Metal in Africa,* exh. cat. New York: African-American Institute, 1982.

d' Azevedo, Warren. *The Artist Archetype in Gola Culture.* Desert Research Institute, Occasional Paper no. 14. Reno: University of Nevada, 1966.

_____, ed. *The Traditional Artist in African Societies.* Bloomington and London: Indiana University Press, International Affairs Center, 1973.

de la Rue, Sidney. *The Land of the Pepper Bird.* New York and London: G.P. Putnam's Sons, 1930.

Donner, Etta Becker. *Hinterland Liberia.* Trans. Winifred Deans. London: Blackie and Sons, Ltd., 1939.

_____. "Kunst und Handwerk in NO-Liberia." *Baessler Archiv* 23, nos. 2-3(1940):45-110.

Eberl-Elber, Ralph. *Westafrikas letztes Rätsel.* Salzburg: Verlag das Burglandbuch, 1936.

Fagaly, William. "The Victor Kiam Collection at the New Orleans Museum of Art." *African Arts* 16, no. 4(1983):28-39.

Fagg, William B. "Two Early Masks from the Dan Tribe in the British Museum." *Man* 55, no.175(1955):161-162.

Fischer, Eberhard. "Die Töpferei bei den Westlichen Dan." *Zeitschrift für Ethnologie* 88, no.l(1963):100-115.

_____. "Künstler der Dan." *Baessler Archiv* n.s. 10, no.2 (1963):161-262.

_____. "Die Bezeichnung der Rollen im Socialsystem der Westlichen Dan." *Zeitschrift für Morphologie und Anthropologie* 60, no.2(1964):242-255.

_____. "Zur Technik des Gelbgusses bei den Westlichen Dan." In *Festschrift Alfred Bühler,* ed. Carl M. Schmitz. Basel: Pharos Verlag, 1965, 93-115.

_____. *Der Wandel ökonomischer Rollen bei den Westlichen Dan in Liberia.* Wiesbaden: Franz Steiner Verlag, 1967.

_____. "Nackenhänger bei den Dan in Liberia." *Baessler Archiv* n.s. 16(1968):99-126.

_____. "Selbstbildnerisches: Porträt und Kopie bei Maskenschnitzern der Dan in Liberia." *Baessler Archiv* n.s.18 (1970):15-41.

_____. "Dan Forest Spirits." *African Arts* 11, no.2 (1978): 16-23,94.

_____. "The Western Guinea Coast: A Short Introduction to West-African Art Province." *Critica d'Arte Africana* 46 (1981):50-65.

_____, and Hans Himmelheber. *The Arts of the Dan in West Africa.* Trans. Anne Buddle. Zürich: Museum Rietberg, 1984. Originally published as *Die Kunst der Dan.* Zürich: Museum Rietberg, 1976.

_____, and George W. Tame-Tabmen. "Erzählgut der Westlichen Dan in Liberia." *Anthropos* 62(1967):686-744.

Gerbrands, Adrianus. *Art as an Element of Culture, Especially in Negro Africa.* Leiden: E.J. Brill, 1952.

_____. "Art as an Element of Culture in Africa." *Anthropology and Art,* ed. Charlotte M. Otten. Garden City, N.J.: The Natural History Press, 1971.

Graburn, Nelson, ed. *Ethnic and Tourist Arts: Cultural Expression from the Fourth World.* Berkeley: University of California Press, 1976.

Harley, George. *Notes,* 1937. Peabody Museum records, Cambridge, Mass.

_____. *Notes on the Poro in Liberia.* Cambridge, Mass.: Papers of the Peabody Museum 19, no.2(1941).

_____. *Masks as Agents of Social Control in Northeast Liberia.* Cambridge, Mass.: Papers of the Peabody Museum 32, no.2 (1950).

Herbert, Eugenia W. *Red Gold of Africa: Copper in Precolonial History and Culture.* Madison: The University of Wisconsin Press, 1984.

Himmelheber, Hans. "Einige Eigentumlichkeiten Westafrikanischer Plastiken." *Ethnologica* 2(1960):407-409

_____. *Negerkunst und Negerkünstler.* Braunschweig: Klinkhardt & Biermann, 1960.

_____. "Die Masken der Guéré im Rahmen der Kunst des oberen Cavally-Gebietes." *Zeitschrift für Ethnologie* 91(1963):216-233.

_____. "Personality and Technique of African Sculptors." In *Technique and Personality.* New York: Museum of Primitive Art, 1963.

_____. "Die Geister und ihre irdischen Verkörperungen als Grundvorstellungen in der Religion der Dan," with Wowoa Tame-Tabmen. *Baessler Archiv* n.s. 12(1964):1-88.

_____. "Gelbgussringe der Guéré (Elfenbeinküste)," *Tribus* n.s. 13(1964):13–24.

_____. "Sculptors and Sculpture of the Dan." In *Proceedings of the First International Congress of Africanists*, ed. L. Brown and M. Crowder. London, 1964, 243–255.

_____. "Die Grossvaterstühlchen der Guéré." *Baessler Archiv* n.s. 13(1965):539–545.

_____. "Le systeme de la religion des Dan." In *Les religions africaines traditionelles*, ed. Placide Periot. Paris: Editions du Seuil, 1965, 75–96.

_____. "Masken der Guéré II." *Zeitschrift für Ethnologie* 91 (1966): 100–108.

_____. "Fälschungen und andere Abweichungen von der traditionellen Kunst in Negerafrika," *Tribus* 16 (1967):15–34.

_____. "The Present Status on Sculptural Art Among the Tribes of the Ivory Coast." In *Essays on the Verbal and Visual Arts. Proceedings of the 1966 Annual Spring meeting of the American Ethnological Society*, ed. June Helm. Seattle: University of Washington Press, 1967, 192–199.

_____. "Dan (Westafrika, Elfenbeinküste) Singmaske 'Gegon' in Maple." In *Encyclopedia Cinematographica*. Göttingen: Institut für den Wissenschaftlichen Film, 1971.

_____. "Dan (Westafrika, Elfenbeinküste) Singmaske 'Polonida' in Glekpleple." In *Encyclopedia Cinematographica*. Göttingen: Institut für den Wissenschaftlichen Film, 1971.

_____. "Dan (Westafrika, Elfenbeinküste) Tanzmaske 'Djaka' in Glekpleple." In *Encyclopedia Cinematographica*. Göttingen: Institut für den Wissenschaftlichen Film, 1971.

_____. "Maskentreiben zum Jahresabschluss in Bianouma, Dan, Elfenbeinküste." In *Encyclopedia Cinematographica*. Göttingen: Institut für den Wissenschaftlichen Film, 1971.

_____. "Das Porträt in der Negerkunst, Bericht über eine Versuchsreihe." *Baessler Archiv* n.s. 20(1972):261–311.

_____. "Einige Miteilungen über Metallmasken der nördlichen Dan, Elfenbeinküste." *Jahrbuch des Museums für Völkerkunde zu Leipzig* 20(1975):131–137.

_____, and Ulrike Himmelheber. *Die Dan: Ein Bauernvolk im Westafrikanichen Urwald*. Stuttgart: Kohlhammer, 1958.

_____, and Ulrike Himmelheber. "Hysterische Erscheinungen bei einem Maskenauftritt der Dan (Elfenbeinküste)." *Ethnologische Zeitschrift Zürich* 2(1972):119–128.

_____, and Wowoa Tame-Tabmen. "Wunkirle, die gastlichste Frau." In *Festschrift Alfred Bühler*, ed. Carl M. Schmitz. Basel: Pharos Verlag, 1965, 171–181.

Holas, Bohumil. *Mission dans l'est Libérien: Resultats démographiques ethnologiques et anthropométriques*. Dakar: Memoires de l'IFAN 14, 1952.

_____. *Craft and Culture in the Ivory Coast*. Abidjan, Ivory Coast, 1968.

Holsoe, Svend E. and Joseph J. Lauer. "Who are the Kran/Guéré and the Gio/Yacouba? Ethnic Identification Along the Liberia-Ivory Coast Border." *African Studies Review* 19, no.1(April 1976):139–149.

Johnson, Barbara C. "Seeking a Name: Four Dan Sculptors of Liberia." M.A. thesis, San Francisco State University.

_____. "Ldamie, Figurative Brass-caster of the Dan." *Iowa Studies in African Art*. Vol. 2. Iowa City: University of Iowa Press, to be published in 1986.

Johnston, Sir Harry. *Liberia*. London: Hutchinson and Co., 1906.

Leuzinger, Elsy. *The Art of Black Africa*. New York and Greenwich, Conn.: Graphic Society Ltd., 1972.

Meauzé, Pierre. *African Art*. Cleveland and New York: The World Pblishing Co., 1968.

Person, Yves. "Les Kissi et Leurs Statuettes de Pierre." *Bulletin de IFAN* 23(1961):1–59.

Riddell, James C. "Labor Migration and Rural Agriculture among the Ghannah Mano of Liberia." Ph.D. diss., University of Oregon, 1970.

Schadler, Karl-Ferdinand. *Afrikanische Kunst*. Munich: Wilhelm Heyne Verlag, 1975.

_____. *Afrikanische Kunst. African Art*. Trans. Charles C. Roberts. Exh. cat. Munich: Stadtsparkasse, 1976.

Schwab, George. *Tribes of the Liberian Hinterland*. Ed. George Way Harley. Cambridge: Papers of the Peabody Museum 31 (1947).

Sibley, James and D. Westermann. *Liberia—Old and New*. Garden City, N.J.: Doubleday, Doran, & Co., Inc., 1928.

Sieber, Roy. *African Textiles and Decorative Arts*. New York: The Museum of Modern Art, 1972.

Siegmann, William C. *Rock of the Ancestors*, with Cynthia E. Schmidt. Suakoko, Liberia: Cuttington University College, 1977.

Tabmen, George W.W., "Gor and Gle: Ancient Structure of Government in the Dan Gia Tribe." Mimeograph, Monrovia, Liberia, 1971.

Tabmen, George Wowoa W. [George Tabmen]. "Death (*ga*) in Dan Culture." *Ethnologische Zeitschrift Zürich* 2(1974):159–182.

Thompson, Robert Farris. *African Art in Motion*. Los Angeles: University of California Press, 1974.

Tulk, Alfred. Journal of trip to Liberia, 1931 and 1933. New Haven, Conn.

Vandenhoute, P.J. *Classification stylistique du masque Dan et Guéré de la Côte d'Ivoire Occidentale*. Leiden: E.J. Brill, 1948.

Wells, Louis T. "The Harley Masks of Northeast Liberia." *African Arts* 10 no.2(1977):22–27, 92.

Zemp, Hugo. "Eine esoterische Überlieferung über den Ursprung der maskierten Stelzentänzer bei den Dan (Elfenbeinküste)." In *Festschrift Alfred Bühler*, ed. Carl M. Schmitz. Basel: Pharos Verlag, 1965, 451–466.